Grow

A 52-Week
Devotional Journal
to Start Each
Sunday

Written by Nancy Umberger

TJS PUBLISHING HOUSE
TJSPUBLISHINGHOUSE.COM

Published in the United States of America

First Edition: September 2025

10 9 8 7 6 5 4 3 2 1

ISBN-13: 978-1-952833-69-4
ISBN-10: 1-952833-69-8

Cover design by TJS Publishing House

Published by TJS Publishing House
www.tjspublishinghouse.com
contact@tjspublishinghouse.com

Dedication

Thank you, Sandie, for inspiring me to take my writing to a different level.

A special thank you to my dad, who, without his heart for the work of the Lord and sharing the gospel, this book would not have happened.

In loving memory of my mom, my prayer warrior, my best friend, my encourager... always pushing me to use my gift of writing.

Introduction

What do you expect from a garden? A harvest? Weeds? Insects that devour your crop? Drought? Too much rain? I believe, if you have ever raised a garden, these things are what you get, whether you ask for them or not. And don't we do our best to rectify the dreadful things that occur in our gardens? We fertilize, pull weeds, smash bugs, and put natural pesticides on our plants, hoping for a bountiful crop.

My question to you is simple. What about your "soul" garden? How do you tend it? Do you expect The Master Gardener to give you a bountiful harvest? Have the weeds of discouragement and despair, anger, and hurt taken over your desire to tend your "soul" garden? These snippets originated out of one person wanting a verse a week... just one verse to think on.

So, when you dive into each week, know that each week is something I have personally experienced. I have walked in the garden with The Master Gardener for a long time, but not until recent years have I allowed The Master Gardener and the Manual to tend my "soul" garden. Although I do not know you personally, I know The Master Gardener knows everything about you, and He is ready to help you tend your "soul" garden. To give you a bountiful harvest. To fulfill dreams. To pull those nasty weeds that have a stronghold on your life. To give you a "hope and a future" (Jeremiah 29:11b NIV).

So, if you are ready for The Master Gardener to begin to prepare the soil of your "soul" garden... then be prepared for extensive digging, sowing of fresh seeds, and lots of weed removal. So, go ahead and dig into the first week. Think about how it can change you, who you can pray for this week, and what area of your "soul" garden needs tending, too.

"Soul" garden blessings,
Nancy Umberger

Sunday Snippet 1

First morning light of the new week. New pages to fill in. We might be writing on the same chapter, but we've been given fresh pages. I was thinking how last week had some "not so good" moments. The scripture I want us to think on is found in Philippians 3:13 NLT "No, dear brothers and sisters, I have not achieved it, but I focus on this one thing: Forgetting the past and looking forward to what lies ahead..."

I don't want to spend my new week looking in the rear-view mirror of last week! I want to look through the windshield of a fresh week...clear and free of bug juice! 🤣 Don't want to be found scribbling and doodling on my fresh pages, nor flipping to past pages! New pages = new beginnings. Happy Sunday! Can't wait for our lunch break! Be safe today. ☕☕

Sunday Snippet Reflections (SSR)

1. How did today's Snippet speak into my "soul" garden?

2. Application to my "soul" garden.

3. One thing I am grateful for:

4. This week, I will commit to pray for:

5. Answered prayers:

Bible/Sermon Notes Date:____________

Scripture:

____

____

Key verses:

____

____

____

Notes:

____

____

____

____

____

____

____

____

____

____

____

____

____

____

Sunday Snippet 2

What a week! Trying desperately not to look in that rear-view mirror. In all honesty, I've glanced one time too many! And then, my friend, I got mad! Mad that someone waltzed in and took what I thought was mine. Something I loved! I love seeds, dirt, and just watching things grow. It's in my DNA (Dirt-N-Awesomeness) 🤣

So, this week, I'm focusing on NOT planting seeds of anger in the garden of my soul.

James 1:20 NLT says, "Human anger does not produce the righteousness God desires." I really desire to produce good fruit in my life...got to move forward...stop looking back...plant a crop of thankfulness...and plow up my rows and rows of anger and hurt!

This week, let's focus on planting new seeds. Seeds of overcoming hurts, seeds of learning a new skill, seeds of helping others, and most of all, seeds of joy. It won't be easy, but our "soul" gardens will be a lot healthier! 🪴🌻💕

When Monday comes, our goal is to plow up the old and start seeding with new seeds. I figure the mean old crows will want to start plucking our new seeds up, so let's make a scarecrow (a reminder of James 1:20) and hang it in our newly planted garden! Trying to stay focused on the new! Change is hard! But thankful you are in my garden, plowing alongside me.

Have a great day! 🚜☕☕

Sunday Snippet Reflections (SSR)

1. How did today's Snippet speak into my "soul" garden?

____

____

____

____

2. Application to my "soul" garden.

____

____

____

3. One thing I am grateful for:

____

____

____

4. This week, I will commit to pray for:

____

____

____

5. Answered prayers:

____

____

____

Bible/Sermon Notes Date:___________

Scripture:

Key verses:

Notes:

Sunday Snippet 3

What a week! I think we did great! 🙂 We both were tending our newly plowed gardens, and I do believe I see new growth in mine. How about you? I've seen, up close and personal, the fruit human anger can produce. Many are harmful to my physical health. I did pull some "bitterness" weeds this past week. I saw they were starting to make their way back, but nope! Can't allow those weeds back.

Today I found a new pack of seeds I'm planning to plant. It's called "Wise Words." It says on the instructions: "Grows and produces fruit within 24 hours. Produces fruit that carries 'Life and death.' Be cautious when picking fruit, as it can have adverse side effects. It can produce love, joy, peace, happiness, and long-lasting growth, or it can produce hatred, hurt, discouragement, pain, grief, loneliness, and stunt growth." Sounds powerful! I hope the fruit I pick produces life.

Words are powerful! So, this package of seeds I plan to sow will bear fruit that either heals or kills. Our verses for the week are:

Proverbs 18:21a NLT

The tongue can bring death or life;....

Proverbs 16:24 NLT

Kind words are like honey—
sweet to the soul and healthy for the body.

I pray we have a good harvest this week. I'm glad I found my "Wise Words" seed package. It's a stark reminder of the power words have, and our Heavenly Father says we can heal or kill with the fruit of our words. Love you! 💕🌱 Let's plan on a huge harvest of flavorful, healing fruit!

☕☕

Sunday Snippet Reflections (SSR)

1. How did today's Snippet speak into my "soul" garden?

____

____

____

____

2. Application to my "soul" garden.

____

____

____

3. One thing I am grateful for:

____

____

____

4. This week, I will commit to pray for:

____

____

____

5. Answered prayers:

____

____

____

Bible/Sermon Notes Date:___________

Scripture:

Key verses:

Notes:

Sunday Snippet 4

I love mornings! Early morning! Before the sun comes up. Why? Because I love to listen to nature wake up. It's the symphony written by the Creator, and all creation is His orchestra.

Mornings. A time to reflect on the day ahead and the days gone by. A time to check my newest section of the garden. The seed packet of Wise Words is flourishing. How's your crop doing? I don't think I killed anyone this week with my words. Probably gave them bouts of food/word poisoning because the fruit that spewed out of my mouth was hurtful, but overall, I was VERY aware of the power of the packet.

I'm looking intently at my soul garden this morning. I'm keenly aware that in the night, critters sneak into my soul garden to steal what doesn't belong to them. It tries to make tending my soul garden more difficult than it should be because that evil critter, in the cover of darkness, is sowing weeds of discord, hurt, lies, pain, hatred, anger, and bitterness between my good fruit-bearing plants.

A new morning has come. Time to reflect. Time to remember that the garden of our soul is fertilized by 4-10-10 (John 4...10:10...Get it? Because John is the 4th Gospel 😆). Here's what the package reads: "The thief's purpose is to steal and kill and destroy. My purpose is to give them a rich and satisfying life."

Let's use 4-10-10 this week! Let's throw THIS fertilizer filled with mercy, grace, peace, and happiness out into the rows of our soul that didn't fare well over the past week because the enemy crept in through the gate of our garden.

Let's spread this fertilizer on our soul garden because it's so rich in nutrients that we can live satisfied! We'll never live free from the weeds that the enemy plants, but I've got a powerful fertilizer that daily enhances my soul garden to become richer and filled with good fruit.

Love you! Enjoy your day! 💕☕💕☕

Sunday Snippet Reflections (SSR)

1. How did today's Snippet speak into my "soul" garden?

2. Application to my "soul" garden.

3. One thing I am grateful for:

4. This week, I will commit to pray for:

5. Answered prayers:

Bible/Sermon Notes Date:___________

Scripture:

Key verses:

Notes:

Sunday Snippet 5

Weariness can settle in if you're trying to keep weeds and critters out of your garden. It's been a busy week! I'm constantly checking my "soul" garden! I DO NOT want those nasty weeds of anger, hurt, sadness, and despair to get a foothold.

I've had to pick from my Wise Word row repeatedly this week. I want my words to be palatable...sweet to the soul...not toxic. I might be in need of planting several more rows of Wise Words. I'll surely let some of the plants go to seed...this way I'll always have seeds on hand.

While I was tending my "soul" garden this week, I noticed a barren plot of land. Why I haven't noticed it before is beyond me. There were no weeds, no grass, no growth of any kind. Just dirt.

I began to think about what I should do with this spot. I started raking over the compacted soil. After what seemed like hours, I began to find garden markers everywhere. You know, the ones that tell you what you have planted in a certain row. I gathered them, placing them in a bucket until I got back to the house.

I dumped them onto the table. My heart ached as I read them one by one. These garden markers were mine! Mine from many years ago...they marked the rows of where I had planted seeds of aspirations, hopes, and dreams. They never grew because I never watered them. It was the plot of regret. I turned each tag over and over in my hands until I wept. All I could say was, "It's too late...just too late to even try to plant any of these seeds again." Tears of regret ran through the crevices of my soul.

You know what? I decided to plow up that little barren spot.

Fertilize it. Put new soil on it and replant the barren plot I'd called "Regret." Why not leave it alone? Well, I was looking through my Gardening Manual, and one chapter stood out to me as I searched for answers on restoring this soil. The chapter was called, "Applying the Isaiah 43:19 NLT Principle." It reads, "For I am about to do something new. See, I have already begun! Do you not see it? I will make a pathway through the wilderness. I will create rivers in the dry wasteland. "

God was already working this barren plot before I found it. He is The Master Gardener...it's no longer a plot of regret. I'm renaming it "Redeeming Rows." He promised to restore the wasteland...I just have to get started. This is the week of NO REGRETS! Just start planting your rows with your aspirations, hopes, and dreams...new ones or old ones...He will supply the rivers to water it. ☕☕

Sunday Snippet Reflections (SSR)

1. How did today's Snippet speak into my "soul" garden?

____
____
____
____

2. Application to my "soul" garden.

____
____
____

3. One thing I am grateful for:

____
____
____

4. This week, I will commit to pray for:

____
____
____

5. Answered prayers:

____
____
____

Bible/Sermon Notes Date:___________

Scripture:

Key verses:

Notes:

Sunday Snippet 6

It's that time of year! Time to rotate the spring and summer crops out to make way for fall and winter crops. The last of the summer sun is beating on my back, salty sweat is dripping from my brow, stinging my eyes, and I have the task of pulling everything up. Whew! It's tiring but rewarding!

As I work pulling up the old, I think about how each row provided food for me, the deer, the box turtle, and my chickens. Never once did the beans, tomatoes, squash, or cucumbers complain about where they were planted, who got more water or fertilizer. Nope, they just grew and gave. With or without extra attention from me, they did their job. The weeds grew and surrounded them, but they grew too! They produced fruit in the midst of often being robbed of nutrients because of the weeds. As I worked my way through the garden, I was convicted that a plant knew more about how to live life than I did, and I'm the gardener!

You see, I've been planted in places and refused to grow. Refused to just take the nutrients The Master Gardener was providing because I wanted the attention He was giving someone else. This attitude stunted my growth and stopped me from bearing the precious, life-giving fruit The Master Gardener designed for me to give away. I chose not to grow. I let the weeds overpower me and stopped growing spiritually. I just refused to grow where He planted me.

This is where The Master Gardener stepped in and had an in-depth conversation with me about cover crops. You may be asking, "What are cover crops?" So glad you asked! Cover crops do specific things. They aid in creating better soil nutrients for the upcoming year, and they help with weed control. I need both for my "soul" garden. I cannot let another year go by where my unattended attitude causes root rot in my soul. So, I've decided to use scarlet clover as my cover crop. It sounds incredibly beautiful!

I grab my bag of scarlet clover seed and head down to the garden. My "soul" garden needs a cover crop, too. Maybe yours is in need of one as well. It's been a hard spring and summer! We've allowed weeds to rob us of what God has wanted to supply. My "soul" garden cover crop is going to be love. I purchased my bag of seeds called Lavish Love from a guy named Peter. He had trouble keeping his love in check for The Master Gardener, too. So, he penned a couple of books to help with this situation.

In his first book, which he calls I Peter chapter 4, verse 8 NLT, he writes "Above all, love each other deeply, because LOVE COVERS a multitude of sins." (Emphasis mine)

Love really does cover a multitude of sins. I believe I'll mix in some "Forever Forgiving" seeds too. Got those from a guy named Paul. His address is Ephesians 4:32 ESV (emphasis mine), should you want some too. I have directions to his place, should you need them. The back of the bag will say, "Be kind to one another, tenderhearted, FORGIVING ONE ANOTHER, as God in Christ forgave you." Use the entire bag! DO NOT SOW SPARINGLY! Sows well with Lavish Love seeds.

Got lots to do! Let's get busy planting our cover crops! Remember, cover crops enrich the soil of our gardens. 💕☕☕💕

Sunday Snippet Reflections (SSR)

1. How did today's Snippet speak into my "soul" garden?

2. Application to my "soul" garden.

3. One thing I am grateful for:

4. This week, I will commit to pray for:

5. Answered prayers:

Bible/Sermon Notes Date:____________

Scripture:

Key verses:

Notes:

Sunday Snippet 7

Weeded✓

Fertilized ✓

Plowed✓

Seeded✓

Watered✓

Now to wait. Wait for the crops to take root and bear fruit. Wait for the cover crop to nourish and refresh the earth. I'm hopeful for a bumper crop! It took a lot of work to get to this point. A lot of sweat and pain and sometimes tears, but here I am. As a gardener, you understand. Understand hard work, sore hands, early mornings, watering when not one drop falls from the heavens, sweating profusely under the relentless rays of the sun, and yet...here we are in a season of waiting. Waiting for our crops to produce and provide exactly what we need in the days ahead.

What do you do in your season of waiting? As for me, I'm already actively planning for my spring crops. Are you? I saw you nod your head in agreement. I have a garden planner. I map out my garden and decide where I will plant my tomatoes, beans, squash, okra...you get the picture.

I'm looking intently at this past spring and the locations I chose for certain crops. I realized I didn't plan appropriately for my sugar peas, and my zucchini needed more space. Okra needs a sunnier place, and I might not need 80-100 tomato plants.🤣 In my planner, I document where I purchased my seeds, fertilizer used, location of where the rows were planted, insect or critter issues, and whether or not I liked the vegetable or fruit.

The season of waiting is also my season of studying. Studying for the next season, when the process will start all over again, means I'll be focusing on gaining knowledge from articles and books, talking to other gardeners, looking through stacks of seed catalogs, ordering seeds, and scratching ideas into my garden planner. Planning helps me look ahead, while reviewing my planner helps me avoid making the same mistakes over and over.

In this season of waiting, I challenge you to take out your spiritual planner. Flip through the pages. Do you see growth? Have you allowed the enemy to destroy your good crops? Are you spending TOO MUCH time just perusing through the catalog (God's Word) and not taking note of any issues? How often are you talking to The Master Gardener? Other gardeners? Searching for specific articles and books? Are you listening to His advice? We study and plan for other areas of our lives. Why not plan for the area that REALLY matters?

Get your spiritual planner out and write this down from 2 Timothy 2:15 AMP (Emphasis mine). "STUDY and do your best to present yourself to God approved, a workman [tested by trial] who has no reason to be ashamed, accurately handling and skillfully teaching the word of truth."

To become more like The Master Gardener, we MUST study the catalog (His Word), find other gardeners (believers) to talk to, and read books and articles so we'll grow and flourish where He has planted us.

I'm challenging us this week to study and plan, but also to

prepare the hardened areas of our hearts so The Master Gardener can begin to sow His seeds within us. 💕☕☕

Sunday Snippet Reflections (SSR)

1. How did today's Snippet speak into my "soul" garden?

2. Application to my "soul" garden.

3. One thing I am grateful for:

4. This week, I will commit to pray for:

5. Answered prayers:

Bible/Sermon Notes Date:___________

Scripture:

Key verses:

Notes:

Sunday Snippet 8

So, how's the waiting going? You're waiting for your new crops and waiting for your harvest. I've checked my gardens religiously for any signs of growth. It seems like it takes a tiny seed SO LONG to grow. I've given all the little seeds everything they need to be successful.

Just this week, while strolling through the rows, there they were. My tiny little beets, spinach, collards, kale, and Swiss chard are all growing. It's SO exciting!

Did you know a seed has to die before it can grow? Isn't that something! Death before life? Here's a seed note (pun intended). The endosperm must die and give up its contents to support life and regenerate the dying seed. If the endosperm doesn't die to support the developing embryo, then new life won't emerge from the dying seed.

I believe this is a hard concept for us to wrap our minds around. Must death really occur before life? Read the above statement aloud. The endosperm MUST DIE AND GIVE UP...not might, not possibly, nope. It must die and then give itself up so it can be more, do more, produce more fruit. That's why the seed dies. It dies in order to produce MORE fruit.

What about you? And me? I found this in The Master Gardener's Manual...

> "I tell you the truth, unless a kernel of wheat is planted in the soil and dies, it remains alone. But its death will produce many new kernels—a plentiful harvest of new lives.

> Those who love their life in this world will lose it. Those who care nothing for their life in this world will keep it for eternity...(John 12:24-25 NLT)

Verse 25 sounds exactly like what happens if the endosperm refuses to give itself up. The seed that refuses to die has nothing to give. It's basically selfish, and its potential ends tragically bearing no fruit. It simply dies.

Look at these two verses carefully. The seeds you and I planted appear to die. They sacrificed everything the second we put them in the ground. Hence, our season of waiting. In truth, the seed HAD to be planted to become what it was ALWAYS meant to be. A tomato seed will not become a tomato plant unless you plant the seed.

Let's look just a little closer and apply this "death before life" concept so we can become ALL The Master Gardener intends. Just like those seeds we tossed in the ground must "lose their life" to fulfill their ultimate purpose (a juicy tomato, green beans, golden zucchini), so you and I must "lose" our worldly life to gain eternity. Simply put, we must be willing to sacrifice everything, just like that endosperm, to follow Christ, grow, and produce much fruit.

As I continued to search for a reason for the 'death before life' concept, I found this statement in The Master Gardener's Manual.

> Then Jesus said to his disciples, "If any of you wants to be my follower, you must give up your own way, take up your cross, and follow me. If you try to hang on to your life, you will lose it. But if you give up your life for my sake, you will save it. And what do you benefit if you gain the whole world but lose your own soul? Is anything worth more than your soul?"... (Matthew 16:24-26 NLT)

In the days ahead, let's choose to die to self to be ALL we were meant to be in the garden of Master. 💕☕☕

Sunday Snippet Reflections (SSR)

1. How did today's Snippet speak into my "soul" garden?

2. Application to my "soul" garden.

3. One thing I am grateful for:

4. This week, I will commit to pray for:

5. Answered prayers:

Bible/Sermon Notes Date:__________

Scripture:

Key verses:

Notes:

Sunday Snippet 9

Walked around the gardens yesterday. Looks like everything is growing. Soon, the harvesting will begin! Until then, I'll wait, watch, and water. How about you? Your garden growing? Seeds doing their job?

My seeds understood their job. 😅 As I strolled through the back garden, I noticed my beets were not looking as good as they did last week. 😩 I love beets! What in the world?! I looked closer. No bugs. No animal tracks. And sadly enough, no water. 😡 The back garden often gets neglected because it's located quite a ways from the water source. Because of the location, it requires extra work to maintain, especially when there's no rain.

I know I'll have to replant and start all over. I will plant them in a different place. I'll give them a better location where the sun shines on them and the water source is close.

I'm sorry I neglected them. Lesson learned! BE AWARE WHERE YOU PLANT! But I'll have another opportunity before it gets too cold to try again.

You know, the garden of our soul often gets neglected. We are SO BUSY, we fail to see the spiritual drought we have allowed to happen. How does this happen? It happens when WE CHOOSE to move away from the water source: the LIVING WATER! We believe, in our minds, we've stored up enough "spiritual fertilizer" (reading the Bible, going to church, talking to other believers...) to last us. But guess what? Fertilizer needs to be "watered in," otherwise, it just sits on the top of the soil surface, so it WILL NOT be enough to BOOST

GROWTH.

"Spiritual" fertilization is no different, my friend. We can't expect to just sprinkle a few verses, a prayer now and then, a good deed here and there, into our "soul" garden and expect it to grow. We have to water it! We have to be close to the water source.

How can we do this if our life is filled with busyness? Busyness often causes our "soul" garden to not flourish because we are allowing the fertilizer to just sit on top of our hearts. We never water it in, because the enemy keeps us so busy! We allow him to cheat us out of the source of water we so desperately need. REMEMBER The Master Gardener never intended for us to experience such a soul drought.

I found some "soul" garden booster plant food in my shed. The bag says Psalm 1:2 & 3. I guess it means one bag with three powerful ingredients. According to the instructions, the ingredients that boost growth are: "delight in the LORD'S instruction and meditate on it day and night." Then, according to this, the third thing is to plant along a riverbank. The Living water flows there, so the results you will see and experience are "bearing fruit each season" and "prospering in all" you do.

You can see the difference almost immediately, but you have to stick with the plan of watering your "soul" garden daily. Maintaining a garden is hard work! So is maintaining the garden of our soul.

But they delight in the law of the LORD,
meditating on it day and night.
They are like trees planted along the riverbank,
bearing fruit each season.

Their leaves never wither,
and they prosper in all they do ... (Psalm 1:2-3 NLT)

Much love! 💕☕💕☕

Sunday Snippet Reflections (SSR)

1. How did today's Snippet speak into my "soul" garden?

2. Application to my "soul" garden.

3. One thing I am grateful for:

4. This week, I will commit to pray for:

5. Answered prayers:

Bible/Sermon Notes Date:___________

Scripture:

Key verses:

Notes:

Sunday Snippet 10

There's nothing more stunning than a garden full of flowers! Colors explode, bees buzz, and sweet smells fill the air. Can you envision it? Can you hear the bees buzzing? Can you smell the sweetness of the air?

I guess, well, I know, I would never say, "There's nothing more stunning than a flower bed full of weeds. Colors fade, bees leave, and that sweet smell is hidden." This is EXACTLY what I'm looking at today!

My vegetable garden is doing great, so I moved my attention to the flower beds. 😩 Where flowers once made their "red carpet" appearance, there is nothing but weeds! Weeds that hinder the beauty from rising. Weeds, like a prison, hold the beauty of the flowers captive. And so I take on the daunting task of pulling weeds. 😪

Seconds turn into minutes and minutes into hours. Exhausted, I keep going. I grab a handful of weeds and notice one lone yellow flower. Hidden, yet there! Amidst the weeds, beauty.

The struggle for this flower must have been hard! Fighting for light, water, and room to expand had to be burdensome. Yet somehow it made it! And now, my little yellow flower has officially made a "red carpet" entrance.🌻

Ever wonder when you'll make your "red carpet" entrance? Ever wonder when someone will come along and pull the weeds out of your "soul" garden so beauty can be seen? Weeds choke, constrain, restrain, hide, and hinder dreams! After much research, I found an article in The Master Gardener's Manual.

He says in Isaiah 61:3 NLT, "To all who mourn in Israel, he will give a crown of beauty for ashes, a joyous blessing instead of mourning, festive praise instead of despair. In their righteousness, they will be like great oaks that the LORD has planted for his own glory."

What does this mean? Well, we're a lot like my sweet yellow flower. It didn't plan on being choked out by weeds, but life happened. The gardener didn't come and pull out the weeds, so growth could occur. We get choked out by the weeds of grief, anger, hurt, despair, discouragement, and a host of other ugly weeds....BUT GOD can and will bring beauty from ashes (a weedy life).

"Beauty from ashes..." What a wonderful reminder that The Master Gardener can take the weeds of our "soul" garden and turn them into "red carpet" moments. The Master Gardener loves to take what once upon a time looked hopeless and turn it into something beautiful!

Beauty from weeds! 🌻 Beauty from ashes! This week, find your "red carpet" moment and allow The Master Gardener to free you from the things that want to hold you captive.☕ 🌻 ☕

Sunday Snippet Reflections (SSR)

1. How did today's Snippet speak into my "soul" garden?

____

____

____

____

2. Application to my "soul" garden.

____

____

____

3. One thing I am grateful for:

____

____

____

4. This week, I will commit to pray for:

____

____

____

5. Answered prayers:

____

____

____

Bible/Sermon Notes Date:___________

Scripture:

Key verses:

Notes:

Sunday Snippet 11

Crisp air. Leaves changing. Breeze blowing. Change coming. Fall has arrived (well, not completely), but the signs are in place and I'm excited! Soups, chili, bonfires, s'mores, all things pumpkin, and time for planning projects for my garden.

My fall garden is looking good! Went out yesterday to check on things. All is good. So, while I wait, I'm planning. I really want to build a high tunnel. It seems to be all the rage these days and seems to work well in our planting zone. So, I'm thinking, why not?

First and foremost, I must consider the cost of the building materials. I've been researching and reading articles for quite some time. Of course, every article, blog, and company tells of only success with their build. I'm currently considering the cost. What I've noticed from every company is that, first, they give you an excellent price. But then, wait for it, they throw in all the tools you'll need for the build. Things like a hoop bender (who just "happens" to have one of these things sitting around? 🤣), a ventilation kit, double-layer greenhouse plastic, and pre-drilled and pre-cut ground posts. My cost just went from $485.00 to $1,220.00. Amazing! Yet, I probably could get around a lot of this by using other materials (just a matter of preference). I need to ask myself, "Do I want it to last or be constantly repairing the tunnel?" In reality, you get what you pay for, so considering the cost means homework and research. So, I'll keep looking.

The Master Gardener has some advice on considering the cost. In The Master's Manual, I found these words in Luke 14:28 NIV: "Suppose one of you wants to build a tower. Won't you first sit down and estimate the cost to see if you have enough money to complete it?"

Even The Master Gardener thinks it's important to consider the cost before jumping into a decision. Especially the decision to follow Him. Just like I'm considering the cost of building a high tunnel, I must consider the cost of following The Master Gardener. "Considering the cost" means careful, detailed, and rational thinking, considering all aspects of what you're getting into before you commit.

There is a cost to follow The Master Gardener, but the harvest is greater than the cost. As I'm sitting here watching the sun wake up the first day of a new week, I'm thankful for the day I considered the cost of following The Master Gardener. It hasn't always been easy! I've had my share of drought and troublesome weeds. Sometimes the harvest was sparse, but The Master Gardener made a way for me to survive. He paid the ultimate price for me and you. He considered the cost and believed we were worth it! He never cheapened the way. He paid the full price!

My question for us in the days ahead is, now that we have "considered the cost," will it cause us to be better Kingdom Builders, or will we cheapen the build by allowing the enemy to sell us tools we don't need? Tools like fear, anger, depression, despair, anxiety, and a host of other things. Don't buy the enemy's tool belt either! Once he gets that wrapped around you, it's not easy to unbuckle.

Yep, today I'm considering the cost. The cost of everything The Master Gardener paid for me and for you. Hope this week brings you joy in knowing you are loved and prayed for!

Much love, my friend! 💕☕☕💕

Working together. Building together.
Praying together.

Sunday Snippet Reflections (SSR)

1. How did today's Snippet speak into my "soul" garden?

    ____

    ____

    ____

    ____

2. Application to my "soul" garden.

    ____

    ____

    ____

3. One thing I am grateful for:

    ____

    ____

    ____

4. This week, I will commit to pray for:

    ____

    ____

    ____

5. Answered prayers:

    ____

    ____

    ____

Bible/Sermon Notes Date:____________

Scripture:

Key verses:

Notes:

Sunday Snippet 12

Well, I thought fall was here, but obviously I'll have to wait.😅 Waiting for change isn't easy. I thought the heat was about over, but I guess even the seasons struggle to change, especially in the south. 🤣 Don't know too many people who embrace change.

I'm still studying my high tunnel build. Got some ideas but nothing is set in concrete. I need to get with the program! Change is coming. Before long, the trees will show off the handiwork of The Master Gardener. Each leaf will be painted with the seasonal strokes of red, orange, and yellow, and we will, once again, stand in awe of this masterpiece. Have you ever noticed that The Master Gardener allows the seasons to change gradually? I believe He does this so we can enjoy the beauty of each season.

I've been thinking a lot about seasons, about change, about life. The trees, flowers, and shrubs never question The Master Gardener when He picks up His Heavenly paint palette and begins to paint the colors of change. Nope, nature accepts the process, even if it might be a little painful. I've never heard a tree complain or say, "I don't want my leaves to be yellow!" or seen a flower lift its leaves up in despair and say, "Why me? Why now? I was just starting to be noticed!"

My question is, why do we get bent out of shape when The Master Gardener decides to change us? It may not be true with you, but I've found I like it when The Master Gardener changes others.😅 I don't experience any pain, but I also don't experience the growth The Master Gardener designed just for me. I'll never become what He planned if I refuse to allow the seed of change to grow. 🌱 Sad, isn't it?

The seasons will change. Nature will make its necessary adjustments as The Master Gardener begins to paint His seasonal masterpiece for all to see!! This week, let's be willing to allow The Master Gardener to take His Heavenly paint palette and change us into something beautiful for all to see! 💕🪴🌻 The Master Gardener tells us in Revelation 21:5a ESV, "And he who was seated on the throne said, 'Behold, I am making all things new.'"

☕☕

Sunday Snippet Reflections (SSR)

1. How did today's Snippet speak into my "soul" garden?

2. Application to my "soul" garden.

3. One thing I am grateful for:

4. This week, I will commit to pray for:

5. Answered prayers:

Bible/Sermon Notes Date:____________

Scripture:

Key verses:

Notes:

Sunday Snippet 13

Don't know about your week, but mine has been crazy! The rain was a welcome sight for my fall/winter garden. Each plant drank its fill of water provided by The Master Gardener. I've got to pick my kale! Nothing tastes better than baby kale, unless, of course, it's ice cream. 🍦 🤣

Quick question. Have you ever planned a project for the day, and two weeks later, you're still working on it? Famous last words: "We should have this done in a weekend." I see you smiling! We've all been there. We are living this dream now. 😅

Two weeks ago, a mighty hawk tried to get my sweet chickens. But the hawk's attempt was spoiled by me running like a crazy lady at it, screaming, "No, no, no! Not today!" All the while swinging a stick. Did I mention I was wearing my favorite rain boots with chickens on them? You should have a good mental picture in your mind by now. 🤣🤣

This is when my "ADD TO THE CHICKEN COOP DIY" weekend project became necessary. Fast forward two weeks, and here we are. Still working. Yesterday was the final push to get it finished.

Let me tell you about yesterday's DIY project. Hardware cloth and a staple gun were about the end of me. In the process of this mighty build yesterday, I got stung twice, the hardware cloth whipped back on me and caught my arm (band-aid moment), and then, wait for it, I was moving the ladder and unbeknownst to me, the hammer was on top of it. And yes, when you move a ladder with a hammer on top, the hammer falls. Whacked me on the back of my head! 😅🤣😂 I wish you could have seen the look on my husband's face when that hammer fell. Priceless!

Are you laughing? Brian and I have laughed and laughed, not because of my injuries, but just because it's so true about the weekend warrior DIY mentality. It always takes us longer. 🤣

Laughing was better than crying! There was no way I could have built this structure alone, nor could Brian. On a serious note, these past two years have been hard for me! And more than likely, you too. One thing I thought about yesterday was, I'm so glad I wasn't working on this project alone! My chickens are happy about this, too! 🐤🐥🐤

It made me realize how thankful I was to have someone else on the project, someone who could measure, cut, and carry a heavy air compressor. God never meant for us to travel through seasons of life (or a crazy DIY weekend warrior build) alone. Even The Master Gardener has something to say about this in Ecclesiastes 4:9, 10 NLT. "Two people are better off than one, for they can help each other succeed. If one person falls, the other can reach out and help. But someone who falls alone is in real trouble."

Today, thank the one who is traveling alongside you! You never know, they may be struggling too. Thank the ones who have stood, and continue to stand, with you in every season. Friendship is two-sided! Never take it for granted! You know, we really DO NEED EACH OTHER!

Our challenge this week is to find those who have stood with us in every season and thank them! We all need a little encouragement to be everything The Master Gardener intends for us to be! Two is better than one. One is a lonely number.

So, I'll start. I'm thankful for you! Thankful for the seasons we've shared and for how The Master Gardener put us together to encourage growth in each other today and in the days to come! 💕☕🚜🌱🌻💕

FYI, I still have to build a door...🤣☕☕

GROW

Sunday Snippet Reflections (SSR)

1. How did today's Snippet speak into my "soul" garden?

____

____

____

____

2. Application to my "soul" garden.

____

____

____

3. One thing I am grateful for:

____

____

____

4. This week, I will commit to pray for:

____

____

____

5. Answered prayers:

____

____

____

Bible/Sermon Notes Date:___________

Scripture:

____

____

Key verses:

____

____

____

Notes:

____

____

____

____

____

____

____

____

____

____

____

____

____

____

Sunday Snippet 14

There's a chill in the air this October morning.✨🍁Finally! Finally, fall is truly here.😊 Of course, in the south, it could be in the 80s in a week. So, I will embrace today!

I've been bouncing back and forth over the concept of a "no till" garden and a tilled garden. If you grew up like me, all we ever knew was a plowed garden. No fancy terminology, no building of layers, no special fertilizers, and no such thing as raised garden beds. We just plowed, made rows, dropped seeds, threw some 10-10-10 and manure in, and worked ourselves into oblivion keeping weeds down to a minimum. Remember those days? 🧑‍🌾🚜 Somehow, we always had a bountiful harvest. Hard work for sure! But rewarding!

A "no till" garden with minimal weeds sounds amazing! But is it? I tried a little experiment this past growing season. I made one garden "no till" and my other garden got the works...tilled, seasoned with multiple fertilizers, manure, compost, and fresh dirt. The "no till" garden received cardboard, compost, dirt, and straw, and raised rows. I worked hard in both garden areas, but the "no till" garden just didn't pull through for me.😪 Or did I not pull through for it?😩

I lost a lot of harvest. A lot of veggies never produced or even sprouted. Why? I believe after much pontificating (I like this word🤣), I failed to truly study the "how-to" of laying out a "no till" garden. So, you might be asking yourself, or you might not be asking yourself, how did I know how to even start this endeavor? Well, I talked to people, bounced ideas around, and even watched a YouTube video, but I never produced the bountiful harvest promised by the videos or the people I spoke to. Something was missing!

At the end of the day, I should have studied! It's okay that I asked people or even watched a video or two, but I SHOULD HAVE STUDIED it myself! Took notes! Researched it deeply, not superficially. Which brings me to an important thought. We, us, you, I CANNOT grow to our greatest potential only off of someone else's knowledge (that's where I failed with the "no till" garden). We must study it for ourselves so we can harvest bountiful crops. Crops of forgiveness, love, peace, friendship, and the list goes on. I learn from others! That's the beauty of life and friendship. Think how much we've learned from each other! And ALL the laughter in between our learning experiences (sometimes we didn't laugh), and we continue to inspire each other to learn! But often, things need a little personal research more than conversation. 😅

I pulled out the Manual. I'm glad I have this Manual in my hands. I realize I don't always run to it like I should. I surely don't study it like I need to, but it has NEVER failed to give me honest direction! Here's what I found for us to think on this week, before we settle into more projects in the garden or take others at their word about our lives: let's vow to study the Manual for ourselves. Let's look deeper into the Manual that The Master Gardener has so freely provided for our growth and knowledge. In Proverbs 3:5-6 NLT it says, "Trust in the Lord with all your heart; do not depend on your own understanding. Seek his will in all you do, and he will show you which path to take."

Let's take our direction from The Master Gardener Himself, and anything else we glean from others along the way is just an additional bonus!

Love you, friend! ❤️💕🚜❤️💕 Have an awesome first day of the week! ☕☕

Sunday Snippet Reflections (SSR)

1. How did today's Snippet speak into my "soul" garden?

____

____

____

____

2. Application to my "soul" garden.

____

____

____

3. One thing I am grateful for:

____

____

____

4. This week, I will commit to pray for:

____

____

____

5. Answered prayers:

____

____

____

Bible/Sermon Notes Date:___________

Scripture:

____

____

Key verses:

____

____

____

Notes:

____

____

____

____

____

____

____

____

____

____

____

____

____

____

Sunday Snippet 15

If you live in NC, you're gonna be running your AC for a long time.🤣 Fall where art thou?! 🍁 🎃 I'm tired of summer allergies in the Fall. Tired of summer temperatures, just ready for consistent weather, but again I live in NC. There's something, though, about anticipating a particular season and only getting a few moments of it. 😩 It's sad! But enough whining.

This week, I was trying an experiment with some students. We were in the garden taking cuttings from tomatoes and peppers, dipping them in rooting powder, and then we put one cutting in water and planted the other in soil. The goal is to see which produces roots first and determine if we can use this product in a hydroponic system.

While in the garden, we found a tomato hornworm. They can and will destroy your crop of tomatoes! This one was covered with tiny white eggs. The tiny eggs were wasp eggs, which were slowly eating the hornworm alive. Good for the Gardener, sad for the hornworm.

Since tomato season is over, I took the hornworm and brushed all the eggs off. You had to have seen that coming.😇 I told my students the eggs could represent sin, fear, hurt, unfairness, anger, loneliness, etc. The list is long! Those wasp eggs were literally killing the hornworm, but he had no way to brush them off or rid himself of the pain UNTIL I intervened.

I brushed each egg off, gave him a little water, made him a home in a plastic bottle, took him home with me, and gave him a chance to become a five-spotted hawk moth.

Isn't this true in life? We all just need THAT SOMEONE to knock off fear, loneliness, and hurt...a friend who is not afraid to see the potential of what you can be! Not afraid to get in the trenches with you and encourage you to be ALL you were meant to be! We need it! And so do others!

You might be wondering whether or not my rescue was worth it. Was it too late? He was pretty lifeless, but I found him yesterday, exactly where I put him, on a fresh tomato plant. Alive! Rescued from pain, given the opportunity to become what God intended.

Today starts a new week. Let's be aware of the potential God has for others and show some grace. Hopefully, we'll receive some grace, too. 😎 You never know, you might be the one God uses to minister to a hurting soul. Remember Proverbs 27:17 NIV which says, "As iron sharpens iron, so one person sharpens another."

☕☕

Sunday Snippet Reflections (SSR)

1. How did today's Snippet speak into my "soul" garden?

____

____

____

____

2. Application to my "soul" garden.

____

____

____

3. One thing I am grateful for:

____

____

____

4. This week, I will commit to pray for:

____

____

____

5. Answered prayers:

____

____

____

Bible/Sermon Notes Date:___________

Scripture:

____

____

Key verses:

____

____

____

Notes:

____

____

____

____

____

____

____

____

____

____

____

____

____

____

Sunday Snippet 16

How are you, my friend? How was your week? I'm beginning to wonder if one week is really more than seven days.🤣This week, I experienced a first. I experienced my birthday without my sweet momma. It wasn't easy! But I found even in the midst of sadness, God always provides joy. Sometimes you have to look for it, but it's always there. ☕💕 New chapters are hard to embrace, you know?

The garden is doing fine. I've got my cover crop planted, and I'm ready to pick some collards and kale. I might try to dust the ground with a few more collard seeds, since NC weather is crazy. 🌿🍁 I just might get a few more greens.

It's crazy to think about, but this year is almost over. It will soon be time to start our seeds indoors, to think about spring, to read the almanac, and dream of the perfect garden.🤣 That's what gardeners do. We DREAM BIG! 🚜🪴🧑‍🌾 But first, we have to plant the tiniest of seeds, snip them, and provide them with enough nutrients to grow. Do you feel the excitement? I do! 🪴🌻 Are you laughing with me or at me? 😂🤣

There's just something about new growth that takes away the hardness that comes from a rough gardening season when your crops didn't produce like you intended. Maybe the scorching sun burnt your plants, or too much rain drowned them out, or pesky pests found their way to nibble on the goods, or like me, you had to replant multiple times and hope something was going to grow. It was a hard growing season this past year. I've come to realize gardeners have to have faith! A lot of it, too!

In a few months, I'll be starting new seeds, praying they will grow, waiting for spring so I can put them in the ground. There's nothing wrong with dreaming big! But, so often, we just sit and dream. Dream about how this or that would be different if only I had....you fill in the blank). I often live in the season of "what if" and "if only."

This week, The Master Gardener pointed me in the direction of a man named Jabez. He was the one who prayed to the God of Israel, “Oh, that you would bless me and expand my territory! Please be with me in all that I do, and keep me from all trouble and pain!” 1 Chronicles 4:9, 10 NLT. And God granted him his request.

Jabez prayed for God to open Kingdom-building opportunities. He simply asked God to bless him: bless him spiritually, physically, materially, and with opportunities to further the Kingdom.

The Master Gardener never intended for you or me to live our days in "what if" or "if only." Seasons come. Some are good and well, some are rough! This is a new day! First day of a new and glorious week! Let's pray like Jabez: for God to take us to new heights spiritually so we can, not only DREAM BIG in the season we are in but DO BIG THINGS with the blessings He bestows on us daily.

Love you! 🤠Happy Sunday! ☕☕

Sunday Snippet Reflections (SSR)

1. How did today's Snippet speak into my "soul" garden?

2. Application to my "soul" garden.

3. One thing I am grateful for:

4. This week, I will commit to pray for:

5. Answered prayers:

Bible/Sermon Notes Date:__________

Scripture:

____

____

Key verses:

____

____

____

Notes:

____

____

____

____

____

____

____

____

____

____

____

____

____

____

Sunday Snippet 17

Oh, the weather outside is...winter? Windy? Frosty? Oops, now we are back to the mid-70s. It will be good to see the sun,🌞 but I'm ready for weather consistency.

I picked some of my kale last night and made a piping hot white bean and kale soup with Asiago cheese bread. It was SO good. 🍲I sure hope you had a great week! Mine was busy but good.

I've been getting all these emails for the new 2022 seed catalogs. That's exciting! Like Christmas in November. 🤣 Thumbing through the online catalogs, there seems to be lots of tools for the gardener, from aprons to gloves to seed planters to new and improved hand tools. It's amazing! And expensive.

I guess I've never really thought about suiting up to go into the garden, but, in reality, I do. I put on boots to protect my feet, gloves to keep me from getting blisters, a hat, sunscreen (when I remember), and sunglasses. All these things work together, so my experience in the garden will be productive. I don't have to have all these items, but more often than not, they're all helpful.

I've really been thinking about this concept of "suiting up" in the garden and how it applies to me spiritually. In Ephesians 6:10-18 NIV, we are told, "Put on the full armor of God, so that you can take your stand against the devil's schemes." I challenge you to look this passage up, read it, write down the pieces of the armor, and then "suit up" for battle...spiritual battle that is. The battle is real!

The devil? Well, he's full of deceit and lies! He's known as the accuser, a destroyer, serpent, dragon, and tempter. His mission is to kill, steal, and destroy. His resume isn't a good one!! 😱

This week, we need to suit up in the armor of God! We need to protect our heart, our mind, carry our shield, use our prayer time as a battle cry for help, and use our voice for worship.

You know, if I'm willing to "suit up" to protect myself from harm in my garden, why shouldn't I be willing to put on the armor The Master Gardener provides for my protection? Let's SUIT UP and be prepared and protected for whatever comes our way in the days ahead. God is faithful!

Love you, friend! ☕🍩💕☕

Sunday Snippet Reflections (SSR)

1. How did today's Snippet speak into my "soul" garden?

____

____

____

____

2. Application to my "soul" garden.

____

____

____

3. One thing I am grateful for:

____

____

____

4. This week, I will commit to pray for:

____

____

____

5. Answered prayers:

____

____

____

Bible/Sermon Notes Date:__________

Scripture:

Key verses:

Notes:

Sunday Snippet 18

It's hard to believe this year is coming to a close. Soon we'll be ringing in the new year. Hard to believe. But first, we have a new week ahead of us, a new page to write on. Hopefully, you've come to the new week with a sharpened pencil to write your God story with. Every day is a new journal entry with The Master Gardener.

Have you ever just sat down and thought about compost? Did you know compost is one of the most important supplements you can give your garden? It's crazy! I've been reading up on it and watching videos. I've tried composting multiple times and never gotten the rich, loamy soil promised. Why? I added all the right ingredients, but after researching, I understand clearly that there were a few things I neglected to do.

Adding the right ingredients is imperative for composting, which I did. It's the next steps that kept me from getting the rich, luscious, loamy soil I desired. I rarely checked on it, kept forgetting to add kitchen scraps, grass clippings, and leaves. I never turned or watered it. The steps I didn't take kept me from seeing the transformation of those useless scraps into a pile of rich, beneficial soil for my garden. Not completing the process kept me from receiving the benefits of composting, which in turn left me with a pile of nothing.

The Master Gardener revealed to me that I should be composting my spiritual life each day. What does this look like? First, you gather the useless scraps (sin) and throw them on the pile. Each day you add to your pile. You add Bible reading, studying, and memorizing the Manual, journaling, and as The

Master Gardener reveals more scraps (sin), you toss them in the pile. Lastly, you turn and water the pile with prayer; lots of it! Prayer changes everything! Prayer is the active ingredient that kick-starts your spiritual compost pile. It adds what is necessary for us to work in unison with The Master Gardener to do what He calls us to do each and every day. We can't afford to neglect any of the composting process. Neglect always costs you something.

Composting our spiritual life is truly a transformation! 2 Corinthians 5:17 NLT says, "This means that anyone who belongs to Christ has become a new person. The old life is gone; a new life has begun!"

I'm working on starting a fresh compost pile for my garden and my spiritual life. I don't want to miss any steps! I'm ready for The Master Gardener to transform my heart and life into something beautiful and useful! Hope you are too!

Happy Sunday! ☕💕☕

Sunday Snippet Reflections (SSR)

1. How did today's Snippet speak into my "soul" garden?

____

____

____

____

2. Application to my "soul" garden.

____

____

____

3. One thing I am grateful for:

____

____

____

4. This week, I will commit to pray for:

____

____

____

5. Answered prayers:

____

____

____

Bible/Sermon Notes Date:___________

Scripture:

Key verses:

Notes:

Sunday Snippet 19

Is it fall? Spring? Winter? Or is summer on your mind? It seems we've experienced all these seasons this past week. I never knew exactly how to dress this week. In NC, you layer to make your multi-season day tolerable. At least you can take off layers as the seasons change throughout the day. 🤣 It's chilly today, and I'm thinking fire pit.

I love fire! I love sitting around the fire and watching the flames turn multiple colors. It's relaxing, and somehow, in those moments, I feel stress-free. No cares. No worries. Just the crackling of the wood burning. Can you hear it? 🍲

Fires don't just happen. It takes work to get a good fire going. Often, after clearing out my garden, I burn the debris. Green, wet plants do not burn well. Remember that. 😆 It's those dry leaves, cedar tree shavings, and small twigs that get the flames roaring and you relaxing.

There are multiple ways to get a fire going: matches, a small torch, rubbing two sticks together, or a ferro rod with a striker. I like a ferro rod. For me, it's a bit of a challenge, but it works. A ferro rod works by striking the rod quickly with a knife or a striker. The exposed pieces are oxidized by the oxygen in the air, resulting in sparks. The sparks light your dry items, like grass, leaves, and small twigs. Then you have a beautiful fire.

Fires are helpful, but we know how destructive they can be, too. This past week, I've sat by the warmth of a fire multiple times, watching the embers burn, feeling the stresses of life just burn away.

As I watch the flames go higher and burn brighter, I'm reminded of the Manual The Master Gardener provided me with. He has a section in the Manual on fire, too. Guess what? Just as a ferro rod and striker can spark a rip-roaring fire, the tongue can do the same thing. WHAT?! 😱 Yep! I just went there! 🤣 Look in your Manual. Go to the section called James. It says:

> In the same way, the tongue is a small thing that makes grand speeches.
>
> But a tiny spark can set a great forest on fire. And among all the parts of the body, the tongue is a flame of fire. It is a whole world of wickedness, corrupting your entire body. It can set your whole life on fire, for it is set on fire by hell itself ... (James 3:5-6 NLT)

I don't want my tongue to be like a ferro rod and striker, starting a spark for potential destruction. I don't want to be responsible for the spark! I've been on the receiving end of "spark throwers", and it's not fun! It's painful, and sometimes you wonder if you'll recover.

The sparks our tongues throw are powerful! The Master Gardener has given us a way to use our sparks for good. It's found in Hebrews 3:13a NIV "But encourage one another daily..." Did you catch that word? DAILY! We all need encouragement! And yes, we need to be the spark of a life-giving fire in other people's lives! Especially in the holiday season. How about this week we use the sparks we throw for good!

Enjoy your day! ☕🍲☕🎶☕

Sunday Snippet Reflections (SSR)

1. How did today's Snippet speak into my "soul" garden?

____

____

____

____

2. Application to my "soul" garden.

____

____

____

3. One thing I am grateful for:

____

____

____

4. This week, I will commit to pray for:

____

____

____

5. Answered prayers:

____

____

____

Bible/Sermon Notes Date:___________

Scripture:

Key verses:

Notes:

Sunday Snippet 20

I really enjoy watching the sun rise and listening to all the sounds of nature. The birds chirp, the owls hoot their goodnights, the rooster crows, alerting everyone to rise and shine, the cows are mooing softly in the distance, and all seems to be in order. A new day is appearing as the sky slowly lights up.

Early yesterday morning, I went hiking and accomplished five miles, with the first half being straight uphill. Once I reached the top, the view was spectacular! But first I had to climb...🥾🥾

The area where I hiked had a huge wildfire back in 2016. The fire burned over 1900 acres! Thankfully, a lot of new growth has happened since 2016. If you didn't know about the fire, well, you might not be able to tell as you sit on the top of Chestnut Knob. If only the forest could talk and share about that horrific day and its awesome comeback of growth.

I was thinking about loss this week. I remembered how my beans just wouldn't grow this past summer. They would come up and then die. One loss after another, but I kept replanting until I finally got a harvest. Loss is hard on a gardener. I put those seeds in the ground expecting a harvest of goodness, but sometimes loss gets in the way, and the gardener is forced to plant again. It's often disheartening! But it sure is rewarding when new growth appears and a harvest is brought in.

We've all experienced a loss or two while gardening. A plant dies, gets a disease, or nothing comes up at all because of bad seeds. But a loss is still a loss. In life, we also experience losses. All losses are painful! Death, friendships, health, you name it.

Any loss is hard! Whether it happens in the garden or the garden of our soul. I believe "soul" garden losses are the hardest.

Have you ever wondered if The Master Gardener cares? Does He have a plan to replant joy and peace, or will He leave you alone? Ever think if your losses can produce a harvest? The answer is YES! The Master Gardener said so in the Manual. Read slowly...

> All praise to God, the Father of our Lord Jesus Christ. God is our merciful Father and the source of all comfort. He comforts us in all our troubles so that we can comfort others. When they are troubled, we will be able to give them the same comfort God has given us ... (2 Corinthians 1:3-4 NLT)

The whole purpose of loss is not only to feel The Master Gardener's comfort, but also to take the harvest of comfort given by Him and give it away. Loss is hard, no matter how you look at it. So, my challenge to you is to take your loss and let The Master Gardener replant your "soul" garden with comfort. Then take the harvest He provides and share it with those who are experiencing a loss in their "soul garden," possibly for the first time. Harvest only happens if we are willing to plant and sometimes replant. Let's be willing to share our harvest. 👩‍🌾🌻🌱🌻

Happy Sunday! ☕☕

Sunday Snippet Reflections (SSR)

1. How did today's Snippet speak into my "soul" garden?

2. Application to my "soul" garden.

3. One thing I am grateful for:

4. This week, I will commit to pray for:

5. Answered prayers:

Bible/Sermon Notes Date:___________

Scripture:

Key verses:

Notes:

Sunday Snippet 21

Hope this morning brings you feelings of relaxation and coffee! ☕☕ Lots of coffee!

What about your week? Good? Bad? Nightmare? Well, I think mine was a mixture. Traffic is nuts, people are near crazy, and I've seen the Grinch up close and personal at least twice! 🤣🤣🤣

With that being said, welcome to the birthday month of Jesus.

We made it through last week ☕☕☕ and now today is the first day of a brand new start. I'm always thankful for Sunday! New page, sharp pencil, fresh outlook, and new adventures.

Speaking of adventures, some little critter has had a grand adventure in my garden. 🐰 I went out to gather some collards, when what to my wondering eyes should appear but leaves nibbled and chewed on. I couldn't figure out how that critter got in because that area is fenced in, so I took a little stroll.

You know what I found? The tiniest piece of fencing was pulled back. Just enough for my critter friend to have a gourmet salad. 🐰🥗

It was only a tiny spot, but the damage was done. I pondered my great salad thief and his remarkable ability to go through such a small space. I guess what my little salad thief saw was an opportunity (ever so small) to crawl through and feast. I'm sure he took a selfie and posted it on his Instagram feed. 🤣

It was such a little hole! I frantically looked in The Master Gardener's Manual for help. You know that Manual has everything! The Manual called it a 'foothold,' which means, according to Webster, is "a firm standing; that on which one may tread or rest securely."

I gave Little Rabbit Foo Foo a foothold, an opportunity. He was treading and most assuredly resting securely in MY garden, all because I had not tended to my fence. I hadn't taken the time to walk the perimeter of my garden to see if anything needed mending because life has thrown a lot at me lately. You know exactly what I'm talking about! Life happens, and things get neglected.

I believe our "soul" gardens get neglected the most. We're too busy checking Facebook, Instagram, YouTube, watching the news, and just being in love with busyness. Should you disagree with the above statement, give yourself a survey and see where you spend the majority of your time. It's alarming how much time we give to these things, isn't it? We're constantly being distracted, which causes us to not check the perimeter of our "soul" garden. Painful truth, isn't it?

Not checking the perimeter of our "soul" garden allows satan an opportunity to creep in and rest securely. What's that old thief nibbling on? He's eating away at your relationship with Jesus. Yep! Look within. Feeling angry, anxious, self-centered, greedy, stressed, mean-spirited, overwhelmed, oversensitive? Guess what? The thief is in your "soul" garden! The Manual says in Ephesians 4:27 ESV, "and give no opportunity to the devil."

How do you mend the fence around your "soul" garden? You check the perimeter of your heart, shore up the holes where you see the devil working, mend it through prayer, and then you get back on track by reading the Manual. Stand guard! The devil only needs the tiniest of holes to destroy the crops you planted and are tending in your “soul” garden.

Funny thing...the devil never has to post on his Instagram feed his invasion of our “soul” garden. We do it for him by our words, actions, and deeds. Let's be better at walking the perimeter of our “soul” garden daily and repairing anything that would give the devil an opportunity to slip in.

Have a great week! 👩‍🌾💕☕ Gotta go check my gardens 🤣

Sunday Snippet Reflections (SSR)

1. How did today's Snippet speak into my "soul" garden?

____

____

____

____

2. Application to my "soul" garden.

____

____

____

3. One thing I am grateful for:

____

____

____

4. This week, I will commit to pray for:

____

____

____

5. Answered prayers:

____

____

____

Bible/Sermon Notes Date:___________

Scripture:

Key verses:

Notes:

Sunday Snippet 22

Can't believe the weekend is over. 😪 Do you ever feel like there's not enough time to do chores, laundry, and enjoy life? Sometimes I just want to enjoy life, without feeling like I'm watching it through a car window, going 70 miles per hour, catching a glimpse here and there. 🚗

I'm thinking about spring gardening. 🌱 🌱 Been thinking about dirt. Yep, dirt has been on my mind this week, because I really desire to plant in soil that will give me my bang for my buck. Seeds are expensive!

The dirt in my lower garden is okay, but it could be better. I need to add some compost. Maybe a tad bit of fertilizer. I also need to cut back some trees so the plants can have more sun. My upper garden is not looking as promising. It's in dire need of amendments to perform at its best.

It's a constant struggle to keep the dirt from growing unnecessary things. I remove weeds and rocks. Otherwise, I'll regret allowing my plants to be choked out by weeds. I really have to work at helping my dirt do its job. It's a daily thing! It's a real thing! 👩‍🌾

I'm sure The Master Gardener has something in the Manual on amending soil. Yes. It does. The Manual had a long section on this very topic. You can find it in Luke 8:4-15, but I'll try to sum it up for us.

The Master Gardener often spoke in parables...which really is an earthly story with a Heavenly meaning. Anyway, back to soil.

As stated in the Manual, a farmer (representing Jesus) planted seeds (the gospel), but unfortunately, not all His seeds landed on good soil. In fact, some of his seeds fell on the path and were trampled, some fell on rocky soil but couldn't sink their little roots deep enough to grow, some seeds fell in the thorns and were choked out, and fortunately, some ended up in good soil and grew.

My question to us: Does The Master Gardener care about the soil of our "soul" garden? ABSOLUTELY! Each soil represents one's spiritual state. The seeds, which represent the gospel, first fell along the pathway and got trampled. This means the seed wasn't received (a hard-hearted person unwilling to receive the gospel seed). They walk away and never experience peace. The second soil was rocky. This person received the gospel seed, but sadly didn't let it take root because religion got hard. Religion seemed to be rules and regulations. The third soil was full of thorns. As soon as the gospel seed started to grow, the person was choked out by the cares of the world. The cares of bills, a job, people, and discouragement. Lastly, the good soil represents a person willing to hear the gospel, receive it, and, although conditions might be tough at times, still produce fruit. This person realizes that a relationship with The Master Gardener is greater than religion. It's not about religion but about a relationship.

You know, I desire good soil for my real garden and my "soul" garden. It takes time to grow spiritually. It's a process. But The Master Gardener has all the amendments I need to grow and produce fruit. Read the parable for yourself this week.

Our verse to think on is James 4:17 NLT which says, "Remember, it is sin to know what you ought to do and then not do it."

Let's keep amending our hearts with the words from the Manual! Have a great week!

Sunday Snippet Reflections (SSR)

1. How did today's Snippet speak into my "soul" garden?

2. Application to my "soul" garden.

3. One thing I am grateful for:

4. This week, I will commit to pray for:

5. Answered prayers:

Bible/Sermon Notes Date:___________

Scripture:

____

____

Key verses:

____

____

____

Notes:

____

____

____

____

____

____

____

____

____

____

____

____

____

____

Sunday Snippet 23

How's every little thing? Feeling rushed? Stressed? Exhausted? I can certainly relate! It's sad when I sit down early Sunday morning and I'm already thinking about Monday. 🤣 Doing this takes away from the day and from my first day of a new week. So, for the moment, I'm going to sit still and enjoy my coffee. ☕☕ Join me! I have a cup just for you. 🤠

Have you ever put something together and ended up with an extra screw, nut, peg, or nail? At the beginning of the school year, I purchased some storage organizers. It didn't look extremely hard to put together, so in my great wisdom, I pushed the directions to the side. What a mistake! Needless to say, I had an extra piece or two, and I had to start over, this time with the directions. 🤣

I've known for a long time that gardening follows the same principle. I often (I should highlight the word often) do things in my garden that feel right to me but are against what I would find in any gardening book. For example, how close things should be planted together, or following all the steps to a "no till" garden. Skipping steps can be costly at harvest time. I speak from experience. Yet, I'm a repeat offender in the world of gardening. I don't mean to be, but I don't always take the time to read what the experts say.

The Master Gardener has a Manual, too. You know, there's something in that Manual for every season in life. I've always had a copy of this Manual in my house, but it doesn't help unless it's in my hands and the words land in the soil of my "soul" garden.

I've asked myself, "What stopped me from searching the Manual? Why did I wait so long to really search it?" The only answer I have is the seasons of my life changed so abruptly I didn't know what to do, and then something inside me said, "The Master

Gardener has a word in His Manual for this season you're walking through." I pulled it off the shelf and started to read, and since that day, I've never been the same. It's the MOST wonderful book I own. I don't always do what it says because my humanness gets in the way, but The Master Gardener is patient and always draws me back to the words in the Manual.

My challenge this week is for us to pull out our Manuals and look for answers to the season we currently find ourselves in. The Manual says,

> All Scripture is God-breathed [given by divine inspiration] and is profitable for instruction, for conviction [of sin], for correction [of error and restoration to obedience], for training in righteousness [learning to live in conformity to God's will, both publicly and privately—behaving honorably with personal integrity and moral courage]; so that the man of God may be complete *and* proficient, outfitted *and* thoroughly equipped for every good work. ... (2 Timothy 3:16-17 AMP)

Let's get into the Manual so when it's time to harvest, we are equipped and ready because we studied and followed The Master Gardener's Manual.

Let's do this together! And encourage each other along the way! 🚜🪴🌻☕☕

Have a great day!

Sunday Snippet Reflections (SSR)

1. How did today's Snippet speak into my "soul" garden?

____

____

____

____

2. Application to my "soul" garden.

____

____

____

3. One thing I am grateful for:

____

____

____

4. This week, I will commit to pray for:

____

____

____

5. Answered prayers:

____

____

____

Bible/Sermon Notes Date:____________

Scripture:

____

____

Key verses:

____

____

____

Notes:

____

____

____

____

____

____

____

____

____

____

____

____

____

____

Sunday Snippet 24

What a year! I peeked in the rear-view mirror of this past year, and WOW! It didn't look so good. Loss, sadness, tears, pain, disappointments, stress, and the list goes on. Then I looked out of the windshield. Big difference! Loss was gain, tears were cleansing, pain turned into purpose, and discouragement was a building block. It was all in how and where I looked. Perspective.

I digress. 🤣 Have you ever heard the saying, "Thank a farmer"? Farmers and gardeners are thanked the least, yet they work so hard to provide for others. I'm thankful I know how to farm and grow veggies! Aren't you? I do it because I love it and it's rewarding. Yes, we know it doesn't come without risk and hard work, yet we are still willing to take the risks. 🚜🪴☕☕🌻

Anyone who knows me will tell you that my favorite people in the biblical account of the birth of Jesus are the shepherds. Why? They were viewed as the least of the least. Never climbed one rung of the social ladder. They just did their job, day in and day out, with zero fanfare, taking sheep from place to place. Taking risks every day.

Think about it. ALL DAY EVERY DAY: sheep. Just you and your sheep. Not ten, but possibly thousands. They knew the shepherd's voice when he called, and he knew the ones who belonged to him. Yet, no fanfare, no thanks from the sheep for leading them to still waters or green pastures. Yet, those sheep, without knowing it, really need a shepherd.

One night, God sent an angel to give the shepherds a personal invitation to see the Messiah born in Bethlehem. Can you believe it? Shepherds were chosen over kings and priests to see this baby.

I often wonder what they felt or what they discussed along the way. I've wondered who took care of their flock while they went to Bethlehem. Maybe the Good Shepherd watched their flock. One thing is for sure: those shepherds didn't let fear grip them. It just put fire in their sandals and a longing in their hearts, and off they went, just like that. No questions. No packing. Just obedience.

Take a minute and look in the rear-view mirror of long ago. See those shepherds? Look as they hurry to the place they were instructed to go. See their faces when they look upon baby Jesus? See those tears trickling down their cheeks and into their beards? Simply tears of joy! They had absolutely nothing to bring to the Messiah but themselves, and you know, that's all He desires.

When they left, each man was changed from the inside out. They worshipped. They told others. They did their shepherding from that day forward with a sense of belonging, for now they belonged to the Shepherd.

As another year is quickly making its debut, let's stop and thank the Shepherd who willingly came with no fanfare, just a baby born in a barn with the mission of rescuing lost sheep. Perspective.

And that, my friend, is why I love the shepherds: fear turned into a fire of obedience. That's my desire as the new year approaches: obedience to the Shepherd.

> So they hurried off and found Mary and Joseph, and the baby, who was lying in the manger. When they had

seen him, they spread the word concerning what had been told them about this child, and all who heard it were amazed at what the shepherds said to them. ... (Luke 2:16-18 NIV)

Sunday Snippet Reflections (SSR)

1. How did today's Snippet speak into my "soul" garden?

2. Application to my "soul" garden.

3. One thing I am grateful for:

4. This week, I will commit to pray for:

5. Answered prayers:

Bible/Sermon Notes Date:____________

Scripture:

____

____

Key verses:

____

____

____

Notes:

____

____

____

____

____

____

____

____

____

____

____

____

____

____

Sunday Snippet 25

Slowly but surely, the high tunnel/greenhouse is happening. I'm so excited just to see the stakes and string marking where it will sit. I envision myself planting away, growing flowers and veggies.

It's funny how I feel about this project. The excitement is beyond words. I've watched "how-to" videos, read blogs, watched DIY channels, and spent countless hours researching exactly what I could afford. And now it's really happening.

Considering the cost was huge! Once I did the research and the math, I found this project was doable. So, what happens now? Now that I've considered the cost and committed myself to this project, I must take the next step.

The next step is to get started. Next steps aren't easy because they often present roadblocks to moving forward. Case in point: I didn't know how to square a building, but there are people who do, and with one phone call and some instructions, it seems my high tunnel/greenhouse is on point. YAY!

The next step is to buy the materials needed to finish this project. I've gone through and written down all the necessary pieces and the price of each item. My goal is to be finished building by the end of January.

Why January? Well, I need to get planting in February for my spring gardens. Envision and plan. I believe one key component to envisioning and planning is writing it down. Draw it out on a board. Put it in front of you daily. It's okay to have a vision, to write it down, and look at it, but you have to take the next step, which is ACTION.

I wondered if The Master Gardener had addressed planning, setting goals, and taking action in the Manual. You know He does! 🤠

> Then the Lord answered me and said,
>
> "Write the vision
> And engrave it plainly on [clay] tablets
> So that the one who reads it will run...."
> (Habakkuk 2:2 AMP)

Writing down a vision/dream is only part of the action required. The second part is doing it, which is often the hardest! I've always said taking the first step is the hardest. So, let's take the first step together.

Ask The Master Gardener to direct you as you write down the dreams/visions for your life, or start with your dreams for this year. Write them down, because writing them keeps them at the forefront of your mind. Post them somewhere so your eyes will see them daily, and you will think on them often. Lastly, begin to flesh them out. ACTION! Don't just look at the dream/vision The Master Gardener has given you and wonder if it will ever happen. Get busy doing it! He will show you where things need to be tweaked or even changed.

This week, let's pick up our pencil and write out our visions/dreams. Post them. Live them. They will come to pass, not without roadblocks and hardships, but let's keep our eyes focused on the Manual and the One who provides all the information we need to live life to its fullest.

Happy planning! 💕☕☕

GROW

Sunday Snippet Reflections (SSR)

1. How did today's Snippet speak into my "soul" garden?

____

____

____

____

2. Application to my "soul" garden.

____

____

____

3. One thing I am grateful for:

____

____

____

4. This week, I will commit to pray for:

____

____

____

5. Answered prayers:

____

____

____

Bible/Sermon Notes Date:___________

Scripture:

____

____

Key verses:

____

____

____

Notes:

____

____

____

____

____

____

____

____

____

____

____

____

____

____

Sunday Snippet 26

Sometimes it's hard not to get discouraged. You with me? The unexpected can be a financial burden. Just when you think you're on the upswing....BAM! Life happens.

I'm here to tell you, when life happens, God provides! He turns burdens to blessings, darkness to light, and panic to peace. It's been that kind of week, so I'm thankful for a new beginning! 😂

The weather hasn't been too cooperative in getting my high tunnel/greenhouse finished. So, while the winds howled, power lines snapped, snow quietly fell, rain rushed in, creeks rose beyond their banks, and the sun came and dried it all up (we had four seasons in one day 🤣), I was thinking about tools. 🤠

Tools are an important part of a gardener's life. There are so many new gadgets to make a gardener's life easier. Some aren't worth the money, and others are worth their weight in gold. I've been thumbing my way through seed catalogs, and every one of them has a section on tools. Each one of them claims to be the latest and greatest to make you a master at gardening. 🧑‍🌾 🪴 🌻

Without the right tools, gardening would be exhausting! Everything would have to be accomplished by hand, or we'd have to be creative and come up with something to help us. I'm guessing that's why seed catalogs offer tools. They know that if you order seeds, you might need the latest and greatest gadget to use in the gardening process. Great marketing tool!

I looked over my tools. They have served me well over the years. Some tools were handed down to me, some were gifts, and others I purchased. I really do have all the things I need to be a successful gardener. Have you ever taken inventory of your tools? Sometimes you come across something and wonder why you purchased it, because it never worked like it did in the commercial. Maybe you've been looking at your tools collecting dust in the corner for months. 🤣🤣

I'm convinced more than ever that taking inventory should be done often. Why? Well, I found things I had broken, some were a tad rusty, many were dull, and some tools were neglected over time. Taking inventory allows us to see EXACTLY what we have and what we need.

The Master Gardener desires that we do a spiritual inventory. He wants us to take inventory of how much time we spend working on our "soul" garden. He wants us to take inventory of how much time we spend reading the Manual. Are we sharpening and caring for the tools He has given us? Or have we just left our spiritual tools in the "soul" garden shed of the past because life happened? Maybe it's time to unlock that shed, drag out the tools that once were your favorite, and take inventory of why you stopped using them. You know, neglected tools can be brought back to life if we clean them up or repair them. They still have purpose! Taking inventory of our "soul" shed will take time, but if it restores the neglected and forgotten tools, it's worth it!

The Master Gardener says it best in 2 Peter 1:3 AMP, "For His divine power has bestowed on us [absolutely] everything necessary for [a dynamic spiritual] life and godliness, through true and personal knowledge of Him who called us by His own glory and excellence."

The Master Gardener has provided everything we need to live victoriously when life happens! The challenge this week is to take a spiritual inventory to see what you have and what you need. It's good for the soul.

Happy Sunday! ☕☕

Sunday Snippet Reflections (SSR)

1. How did today's Snippet speak into my "soul" garden?

____
____
____
____

2. Application to my "soul" garden.

____
____
____

3. One thing I am grateful for:

____
____
____

4. This week, I will commit to pray for:

____
____
____

5. Answered prayers:

____
____
____

Bible/Sermon Notes Date:___________

Scripture:

Key verses:

Notes:

Sunday Snippet 27

I love Sunday Snippet day! It's like fresh linens, the first snow, that new car smell, crisp air, first signs of spring. It just represents a fresh, beautiful first day to a new God-given week. We were blessed to wake up to this day! There will never be another one like it.

As a gardener, you and I both know the importance of being prepared. We prepare for storms, for frost, for plant-eating insects, and for drought. We get our tools prepared for use, and our layout and plans for our gardens are prepared. We're ready because we planned ahead. That's what gardeners do. 🧑‍🌾 We prepare without knowing when a storm will come, or if a drought will crack open the ground, or if insects will snack on our favorite veggies. But we're ready! 😆

This week, for me, (dramatic music 🎶) it was READY, SET, (now imagine the sound of screeching brakes)...STOP. What I was preparing to do came to a complete stop due to circumstances beyond my control.

I can't say that I'm overjoyed with the way things worked out, but when the right time comes, I'm prepared! I'm ready!

In times of planning and preparation, I look to the Manual. You know, there's a lot in the Manual about preparing and being prepared. God opened the Red Sea, and a dry road was prepared for the Israelites to cross. Noah prepared a boat that kept his family and animals afloat for a LONG time. God prepared a three-night stay for Jonah inside a great fish. David prepared before he loaded up his sling by locating five smooth stones. Rahab prepared a window by hanging a red cord from it for the safe rescue of herself and her family. John the Baptist prepared the way for the coming of Christ, and Christ prepared a way for us to spend eternity with Him.

There's much to be said about being prepared! The Master Gardener explains:

> "Therefore, stay awake, for you do not know on what day your Lord is coming. But know this, that if the master of the house had known in what part of the night the thief was coming, he would have stayed awake and would not have let his house be broken into. Therefore you also must be ready, for the Son of Man is coming at an hour you do not expect." ... (Matthew 24:42-44 ESV)

Being prepared for this day is easier than preparing for a season of gardening. I've learned over the years that it's simply about a relationship with Jesus, not a religion, which has prepared me for my eternal destination. I'm prepared for this day. I'm ready.

My challenge this week is to live ready for His return. It could be today, and what a glorious day that would be! 💕 ☕☕

Sunday Snippet Reflections (SSR)

1. How did today's Snippet speak into my "soul" garden?

2. Application to my "soul" garden.

3. One thing I am grateful for:

4. This week, I will commit to pray for:

5. Answered prayers:

Bible/Sermon Notes Date:___________

Scripture:

Key verses:

Notes:

Sunday Snippet 28

What did you prepare for last week? Anything? I do believe everyone in the surrounding area is prepared for the "winter" storm, 🌨 because there's no bread, milk, or eggs at the grocery store. People from the north don't understand that sentence, but the very word "snow" throws southerners in a panic. 🤣

Needless to say, the high tunnel/greenhouse hasn't been worked on. Ice and snow stopped the project. In fact, ice and snow stopped a lot of other things, too, like school, for a couple of days. 😅 I admit it was nice.

A lot of my teacher friends took the break to organize closets, drawers, pantries, and to deep clean. I thought about doing the same, 🤣 but I didn't. In fact, I chose to be still. To rest. ☕ Sometimes you don't know how tired you are mentally and physically until you stop. You know what I mean?

At first, I felt bad for not organizing or pitching stuff out because often, without knowing, people make you feel bad for the decisions you make. As I write this Sunday Snippet, I feel better for taking the time to be still. I have a clearer picture of my calling and what God desires from me.

Being still is NOT usually in my vocabulary, but it was time. Time to allow myself to feel the emptiness of no longer having my mom, time to reflect on what steps God is calling me to do and begin to prepare my heart and mind, time to remember the many blessings I've received in the midst of a sorrowful season, and time to renew my commitment to my Heavenly Father. So, all was not lost. 🧹 😂

Let me ask you, dear friend, when was the last time you sat with The Master Gardener and chose to be still? Still enough to hear Him whisper? Still enough to feel His presence? Sitting still in the presence of The Master Gardener promotes healing from our deepest wounds, gives us a fresh fire in our souls, renews, and revives the dreams He planted in our "soul" garden. It's really important to learn to embrace moments of being still.

Being still for me is often awkward. I love to talk and go, 🤣 but sometimes The Master Gardener intends for me to stop, to listen, to come into His presence with no agenda except to sit beside Him and be still. To be still and KNOW (not hope) He is God and desires to speak into my life. In Psalm 46:10a NIV He made that clear where He instructed me. "He says, 'Be still, and know that I am God;...'"

As the sun rises on this Sunday morning, I pray that you and I will take time to be intentional and "be still and know" that He alone is God! And He has glorious plans for our lives!

☕🍩☕🍩 (And yes, it includes doughnuts and coffee.)

Sunday Snippet Reflections (SSR)

1. How did today's Snippet speak into my "soul" garden?

2. Application to my "soul" garden.

3. One thing I am grateful for:

4. This week, I will commit to pray for:

5. Answered prayers:

Bible/Sermon Notes Date:___________

Scripture:

Key verses:

Notes:

Sunday Snippet 29

Once again, a wintry, icy weekend. Not enough snow to sled or build a snowman, but enough icing on the road to make it a little dangerous. And baby, it's cold outside! ❄️🧊❄️

So, on this cold day, I found myself looking at seed catalogs 🤣 and ordering more seeds to add to my collection. Have you ever looked at a seed catalog? The pictures of what I have the potential to grow in my garden are flawless. I see mouthwatering tomatoes, luscious berries, beautiful broccoli, perfect melons, and flowers for all occasions. I dream of a garden where I produce crops with such perfection and beauty.

Those seed catalogs I look at are made - not only for ordering seeds - but to make me desirous to grow the perfect veggies, fruit, and flowers. I order seeds, based not only on knowledge, but more often than not, the picture lures me to dream of this plant in my garden. It's true, a picture is worth a thousand words. The image speaks deeply to the gardener in my soul, and then I order in hopes of sharing my own photos of picture-perfect veggies. 📷

All these beautiful, flawless pictures got me to thinking about myself and the image I portray to others. When people look at me and what I profess to be, do they see me or a photoshopped version of Jesus? I don't want to be a photoshopped Christian! Following Christ isn't easy! He never said it would be!

I desire the picture I portray of Jesus to make someone want to know Him. Looking at the Manual, I found one man whose life was so transformed that he told people to "follow his example." Kinda bold, don't you think? Philippians 3:17 NIV says, "Join together in following my example, brothers and sisters, and just as you have us as a model, keep your eyes on those who live as we do..."

Paul was the real deal! He wasn't perfect, but he lived a life in front of others, so that when they saw him, they saw a picture of Jesus.

Can we say to others, "follow my example?" Paul even encourages his readers to look for others who lived like him. In thinking about our lives, I'll leave you with these thoughts:

A picture IS worth a thousand words, so what kind of picture are you to others? How do people see The Master Gardener in your life? Can you say, "follow my example?"

It's never too late to be all The Master Gardener intends for us to be! When I look at seed catalogs and the images of all the plants, I begin to desire for my garden to look the same. So, I encourage you to look in the Manual and see the potential you have to grow! To grow into the image of Christ. My deepest desire is to grow into a better image of Jesus. How about you?

We made it to a new week! Whew! The sunshine was such a blessing! 🌞 Even though it was still cold, the sun gave me a mental and physical boost (one I needed). How about you?

Yesterday, I was finally able to get all the things needed to build the framework of my high tunnel/greenhouse. Home Depot didn't sell the plastic I needed, but it's on the way. 😆

After several hours, the high tunnel/greenhouse is sitting in its new home. Just a shell, unfinished until next weekend, but nonetheless, I can see the empty structure. I cannot wait to fill it up! I stood back and envisioned myself planting, growing things from start to finish in my greenhouse, and then planting all the goodness in my garden. I'm just so excited! Although it's just the framework, I could still envision the potential of what could happen once completed.

Do you dream like this? You see the potential of land, or a house, a job, a garden, or even the framework of a high tunnel/greenhouse. Ever notice how elated and joyful you are in these dreams? It's good to be refreshed with new ideas, dreams, and projects. It's like a salve to a weary soul.

I was thumbing through the Manual this week and realized The Master Gardener likes to start with just the framework, too. You know, He was a carpenter. 😂 In Psalm 139:13 NLT it says, "You made all the delicate, inner parts of my body..." In Psalm 139:16 NLT we are reminded, "You saw me before I was born. Every day of my life was recorded in your book. Every moment was laid out before a single day had passed."

The Master Gardener started with the framework of our lives before we were ever conceived. He saw potential! He envisioned ALL the wonderful things He wanted to do for and in us. He desired to fill our lives with things that would produce good fruit. Remember, The Master Gardener has a plan. Jeremiah 29:11 NLT says, "For I know the plans I have for you," says the Lord. "They are plans for good and not for disaster, to give you a future and a hope."

The realization of it all is The Master Gardener never pushes His plan on us. He gives us the ability to choose how we fill our "soul" garden. In this new week ahead, let's allow The Master Gardener to take inventory of what WE have chosen to fill our life with and say to Him, "Clean out the greenhouse of my "soul" garden...start something new and fresh inside me so the fruit I bear will bring honor and glory to You."

Are you willing to let The Master Gardener fulfill the potential He has for your life? I am! I'm ready for my "soul" garden to flourish with the fruit He desires.

Happy "soul" garden cleaning day! 🪴🪴☕☕

Sunday Snippet Reflections (SSR)

1. How did today's Snippet speak into my "soul" garden?

2. Application to my "soul" garden.

3. One thing I am grateful for:

4. This week, I will commit to pray for:

5. Answered prayers:

Bible/Sermon Notes Date:___________

Scripture:

____

____

Key verses:

____

____

____

Notes:

____

____

____

____

____

____

____

____

____

____

____

____

____

____

Sunday Snippet 30

This week! 🤔
I really don't need to elaborate. It seems that those two words give us a clear understanding of what our week has been like. I'm thankful for another Sunday (a fresh start). No need to look in the rear-view mirror of last week. It's time to move forward.

High tunnel/greenhouse update: plastic is on part of the structure, but we ran into a small issue. I need to think it over or find a genius! 🤣 It's beautiful so far.

I stood inside my high tunnel/greenhouse. It felt so good! As I was standing there thinking about all the seeds I could plant, I thought about all the other gardeners who were thinking about the same thing. 😆 Comical yet true. We're all dreaming!

It's the Season of Planting. Do you know what you are going to plant? Have you had time to plan out your garden? I get it! Life gets in the way, and then the Season of Planting is here, and you go into overdrive trying to get started and just hoping to plant something.

I looked over my seed packets yesterday. I'm working on how much of each to plant and exactly where to plant them. It's an exciting time, especially with a high tunnel/greenhouse to start my seeds in.

I have LOTS of seed packets. Seed catalogs are my weakness because I see what the seed has the potential to become, and just like that, I order it. I'm guessing there are worse addictions than ordering seeds. 🤣

What if I took one seed from each packet and planted it?

That would be crazy! There wouldn't be a harvest of any kind. In fact, I'm not guaranteed the one seed I planted would even come up. Then I've lost the whole crop I had planned on. Nope, I'm planting lots of seeds because I desire a bumper crop at harvest time.

I realize that sounded crazy, but the Manual speaks of how we should sow seeds. Really! It does! 2 Corinthians 9:6 NLT tells us, "Remember this — a farmer who plants only a few seeds will get a small crop. But the one who plants generously will get a generous crop."

Isn't this the truth! I realize this verse is about giving. But think about what we have to give. We may not have lots of money to give, but we can sow seeds (emphasis on seeds) of kindness, encouragement, helpfulness, listening, and the list goes on. Your Season of Planting will not look exactly like mine, but we should all be striving for a bumper crop. One seed isn't enough.

This week, let's take inventory of our "soul" garden seed packets. Let's look and see what we can start sowing in order to get the harvest The Master Gardener desires for us to gain a bumper crop. Together, let's plan on sowing seeds that will bring honor and glory to The Master Gardener! He promises to give the increase. We have to be willing to sow the seeds. It's our Season of Planting! 🪴🪴☕☕

Sunday Snippet Reflections (SSR)

1. How did today's Snippet speak into my "soul" garden?

2. Application to my "soul" garden.

3. One thing I am grateful for:

4. This week, I will commit to pray for:

5. Answered prayers:

Bible/Sermon Notes Date:___________

Scripture:

Key verses:

Notes:

Sunday Snippet 31

What will be your attitude on this first day of a new week? Attitude is everything! We choose it. As Friday, the 18th, marked one year since my precious momma entered the presence of her Heavenly Father, I made an important decision. I decided to have an attitude of celebration amongst my rubble of grief. I celebrated how much she loved you and me, her prayer life, her love for the scripture, and her love for her Savior. Attitude is everything!

So, this leads me to my high tunnel/greenhouse update. It's finally finished. 🧑‍🌾

This past week, as you know, the wind picked up and we felt its power. This was truly a test for my high tunnel/greenhouse. Would it be able to withstand the high winds that howled around it (and at that point through it) since it wasn't finished? 🤣

I worried through the night. I wondered if, when I got up, the plastic would be hanging from the trees. 🤣 Thankfully, it withstood the wind and the rain. 🌬️🌧️

I did some serious pontificating on this matter.

I came to this conclusion. The only reason the plastic remained on was that it was fastened securely to its anchor. Now this got me to thinking. What am I anchored to when the winds of weariness howl within my soul, the rains of regret pelt on the windowpane of my heart, and unexpected storms swirl in my spirit?

Immediately, I ran to the Manual! I had to find out if The Master Gardener had a way for me to address these elements of life. Here's what I learned. Isaiah 41:10 NLT says, "Don't be afraid, for I am with you. Don't be discouraged, for I am your God. I will strengthen you and help you. I will hold you up with my victorious right hand." We learn in Exodus 14:14 NIV "The LORD will fight for you; you need only to be still." Mark 4:37, 39 NLT teaches, "But soon a fierce storm came up. High waves were breaking into the boat, and it began to fill with water... When Jesus woke up, he rebuked the wind and said to the waves, 'Silence! Be still!' Suddenly the wind stopped, and there was a great calm."

In this life, storms will come! Some will feel like a spring shower, and others will come with hurricane force, but dear friend, storms reveal where we have placed our anchor. Storms reveal the depth of our root system. You may be struggling. The winds may be howling around your "soul" garden, but take heart! The Master Gardener still has the power to speak over your storm, and guess what? When He speaks, the storm listens. Hallelujah!

Put your roots deep this week, my friend! 🪴🌻☕☕

Sunday Snippet Reflections (SSR)

1. How did today's Snippet speak into my "soul" garden?

____

____

____

____

2. Application to my "soul" garden.

____

____

____

3. One thing I am grateful for:

____

____

____

4. This week, I will commit to pray for:

____

____

____

5. Answered prayers:

____

____

____

Bible/Sermon Notes Date:____________

Scripture:

____
____

Key verses:

____
____
____

Notes:

Sunday Snippet 32

Rainy days. You either love them or you don't. I don't mind the rain. It's just painful when it traps me inside day after day, but I have a little high tunnel/greenhouse to go into and stay dry. I'm loving it!

I've started SO many seeds. Saturday, we chopped down trees so the lower garden could be extended (that way I can grow more veggies). 🍅🌶️🫑🥔It's always such an invigorating time of year! Seed starting brings excitement to what lies ahead.

If you came to see my high tunnel/greenhouse, you'd see that everything I've planted has a label. I know exactly what tomatoes I have, what kind of peppers I planted, and the dates I planted them. I plan to start more seeds this week. 🧑‍🌾 Yay!

I watch a lot of gardening videos! I mean A LOT! I really enjoy learning new things and trying different techniques. I recently watched one where the guy was saying it wasn't necessary to label what you plant. WHAT! That's crazy! He then said once his seedlings grew, he knew exactly what they were. I do know the difference between a tomato and a pepper, but I plant a lot of varieties, so I label things. He must be a garden genius. 🤣

However, this did make me think about my life as a believer. I thought about the things I allow to be planted in my "soul" garden. I thought long and hard about what that particular gardener said: "I know what I planted, so I don't need to label anything, because I'll know by the fruit."

Ouch! Do we know what we've allowed to be planted in our "soul" garden? We often attach labels to our "soul" garden to help others identify our gifts and abilities or to draw attention to the woes of our lives. For instance: prayer warrior, kind-hearted, willing to serve, great singer, devoted to friends and family. You get the picture. It seems we want everyone to see and read our label. Labels scream, "Look at me! See what I'm doing. Do you approve? Or disapprove?" Think about this for a few minutes. We're all guilty of labeling things in our lives. We even put negative labels inside our "soul" garden. For instance: Rough past, hurt by others, trapped, grief-stricken, lonely, I'll never make it, and so on.

These labels are devastating to the growth of good fruit. The labels we put on ourselves aren't always positive, but labels are real. The Manual sheds a different light on this matter. Matthew 7:16 NLT says, "You can identify them by their fruit, that is, by the way they act. Can you pick grapes from thornbushes, or figs from thistles?"

Fruits were a common metaphor in both the Old and New Testaments. They represented the outward manifestation of a person's faith, their behavior, and their works. I know this verse sits in the midst of understanding how to identify a false prophet, but anytime we live outside the will and word of God, we are falsely representing God.

We shouldn't need to put spiritual labels in our "soul" garden for others to read. People should simply see our fruit and be able to identify us as those who belong to Jesus. Even in our most trying season of life, our fruit should be so evident that no one should ever have to read our handwritten label to know we belong to The Master Gardener. My friend, our label was etched on our lives the day we accepted Him to rule over our "soul" gardens. Our label simply reads 'FORGIVEN'. Let's allow this label to identify us from now on! ☕☕

Sunday Snippet Reflections (SSR)

1. How did today's Snippet speak into my "soul" garden?

____

____

____

____

2. Application to my "soul" garden.

____

____

____

3. One thing I am grateful for:

____

____

____

4. This week, I will commit to pray for:

____

____

____

5. Answered prayers:

____

____

____

Bible/Sermon Notes Date:___________

Scripture:

Key verses:

Notes:

Sunday Snippet 33

Talk to me about your week. Was it easy? Rough? Did everything go your way? Did you experience any rough patches in the coming and going of your day?

This past week has been very hard for me. Too much to tell. There's no emoji to capture it, but a few words to describe it would be: loss, hurt, empty, uncertainty, disappointment, and lonely. You know the feeling. People all around, but you still feel alone.

You might be thinking, "What in the world does this have to do with gardening?" Well, I've been pondering on this thought, "the silence of a seed." Have you ever wondered what a seed would say if it could talk? 😂

Maybe it would say, "Don't cover me with dirt." Or "I need a drink of water!" Better yet, "It's dark under here." 🤣 Nope, a seed is silent, yet in its silence it's working. Developing. Growing. In fact, it's literally dying in order to reach its full potential, and it never complains. Never complains to its Maker that it got too much water or not enough, the sun was too hot or that the last cold snap just about finished it off, or that the dirt it was planted in doesn't contain the nutrients it needs to thrive or even survive. Nope. A seed is silent. The seed just perseveres, often in the midst of the most adverse situations.

The thing about a seed is that you cannot see the struggle it goes through to break out of the ground and grow. All we see is the tiny seedling that appears, which turns into a much stronger plant and then produces fruit. The process that happens beneath the dirt is hidden. It's unseen by the gardener.

You and I are a lot like seeds. Often we feel buried by the dirt of worry, anxiety, loneliness, change, uncertainty, and even grief, but The Master Gardener knows and He sees beneath the piles of dirt. He sees our potential. We just have to persevere through one more rainy season, another few weeks of heat, or a surprise cold snap. Just persevere until we break free and rise up out of the dirt.

Grab your Manual. Turn to Genesis 16:13 NIV **You may want to read all of this chapter.** Verse 13 says, "She gave this name to the LORD who spoke to her: 'You are the God who sees me'..." Hagar understood the feeling of being buried beneath the dirt of fear, loneliness, worry, and mistreatment. ("The God who sees me" translated in Hebrew is the name El Roi) El Roi was there. He never took His eyes off of her.

He sees you!
He sees me!

El Roi. Say it softly. Let it sink into your "soul" garden. Isn't it powerful! What a beautiful name of God. When we feel lonely, invisible, and even forgotten, we're not! The Master Gardener sees everything. He sees our struggle to rise up out of the dirt! He's aware of everything that is happening beneath the dirt. He is El Roi.

This week, say this name as often as you need it as a personal reminder that "He sees me." You are not forgotten, invisible, or lost. El Roi sees you and He knows exactly where you are planted, and He is watching your struggle, yet all the while providing the strength for you to rise up and produce fruit.

What a beautiful name...El Roi! ☕☕

Sunday Snippet Reflections (SSR)

1. How did today's Snippet speak into my "soul" garden?

____
____
____
____

2. Application to my "soul" garden.

____
____
____

3. One thing I am grateful for:

____
____
____

4. This week, I will commit to pray for:

____
____
____

5. Answered prayers:

____
____
____

Bible/Sermon Notes Date:___________

Scripture:

____

____

Key verses:

____

____

____

Notes:

____

____

____

____

____

____

____

____

____

____

____

____

____

____

Sunday Snippet 34

Another week has gone. Everything that happened is now just a memory. Isn't that something? Those moments, whether good or bad, will never pass our way again. Yet some of those moments will be etched in our minds forever. That's a sobering thought.

I don't know what kind of week you had. Hopefully, it was good and very productive. For me, my week was hit once again with loss and sadness. I'm tired of death, funerals, loss, pain, and the tidal wave of emotions that follow. I'm glad Sunday is on the calendar and a new day is on the horizon. Aren't you?

Can you believe this weather? This week has been crazy! Sunny. Rain. Thunder. Lightening. Snow. Wind. Temps in the teens. I really don't enjoy the flip-flopping of the weather, especially when I have tender seedlings trying to establish themselves. My mad dash yesterday to bring everything inside from the high tunnel/greenhouse and to cover all my blueberry bushes looked like a whirlwind. I didn't get everything covered, but I tried. 😪

Speaking of crazy weather patterns, as you know, plants don't like it! They truly cannot withstand one extreme to another without assistance from the gardener. When weather patterns fluctuate drastically, the plant struggles to perform and grow. If it's too hot, a plant wilts. Too cold, it dies. Consistent weather conditions are what a plant longs for, but a plant has to have a gardener to ensure those conditions can be met.

In the wake of going to two funerals back to back, it has caused me to think about the brevity of my own life. The Manual describes life as vapor.

> Yet you do not know [the least thing] about what may happen in your life tomorrow. [What is secure in your life?] You are *merely* a vapor [like a puff of smoke or a wisp of steam from a cooking pot] that is visible for a little while and then vanishes [into thin air]. ... (James 4:14 AMP)

I want my life to count for Christ! Just like a plant, I often struggle with being in a good spiritual environment consistently. I talked to The Master Gardener regarding the spiritual temperatures a believer faces, and this is what He revealed.

God sent a message to the church people in Laodicea. (Revelation 3:15-22) He told them He knew everything they did, and His desire for them was to either be hot or cold, but being lukewarm wasn't an option. In other words, don't straddle a fence when it comes to your relationship with The Master Gardener. Don't compromise! I truly desire consistency in the storms I weather. I don't want to be wishy-washy when life's weather conditions fluctuate or change on a dime. I want to be consistent in my love and service to Him! I want to be faithful!

The Master Gardener isn't asking either one of us for perfection. He's asking us to be consistent in our "soul" garden when the temperatures of life fluctuate or completely change. Even when we experience multiple hardships in one day, we shouldn't compromise.

Consistency is the key that opens the door for The Master Gardener to bless us here and in eternity. Eternity, my friend, is everything!

Make spending time with The Master Gardener a priority this week. Let's not be lukewarm in our approach to the Manual or to The Master Gardener. Let's make every moment count, knowing we cannot make it without the assistance of The Master Gardener's hand.

Sunday Snippet Reflections (SSR)

1. How did today's Snippet speak into my "soul" garden?

2. Application to my "soul" garden.

3. One thing I am grateful for:

4. This week, I will commit to pray for:

5. Answered prayers:

Bible/Sermon Notes Date:___________

Scripture:

____

____

Key verses:

____

____

____

Notes:

____

____

____

____

____

____

____

____

____

____

____

____

____

____

Sunday Snippet 35

If you're reading this, it means you survived another week. 🤣 We pulled out all the survival knowledge we had and put it into action. From handling life to helping people, we survived! 🤣 Today we rest, if only for a moment. Rest.

I'm enjoying my high tunnel/greenhouse! Of course, with no way to regulate temperature, sometimes my plants get a little too hot. Heat is great for germination, but not so great for the seedlings that have popped up out of the dirt. Especially my cool-crop friends.

I really have to watch over my plants, making sure they don't wilt, providing proper air flow through the high tunnel/greenhouse, watering all plants on a regular basis, and possibly moving some to a better environment. I really want my endeavors to be successful. It's an exciting time for me. I love gardening! 🚜 It's a great outlet for reducing stress.

Just as my plants are dependent on me to provide support in adverse situations, I also rely on The Master Gardener for support when I'm going through difficult times. Think about it. If you and I were plants, we would desire to be in a prime location. A place where we can receive the right amount of sun and water to grow. A place where someone would keep us from wilting and move us to a better location to help us survive and thrive. We would long for a place where we could produce lots of fruit. Right? Remember, our care comes from the watchful eye of The Master Gardener — the One who sees the signs and then provides exactly what the plant (which is you and me) needs at the exact time. The Master Gardener's timing is always right! Even though we might question His choice of heat (pressures and stress), not feeling like we are experiencing enough rain (blessings), The Master Gardener knows exactly how to grow us so we will produce an abundance of fruit.

The Manual says this:

> "But blessed is the one who trusts in the Lord,
> whose confidence is in him.
> They will be like a tree planted by the water
> that sends out its roots by the stream.
> It does not fear when heat comes;
> its leaves are always green.
> It has no worries in a year of drought
> and never fails to bear fruit." ... (Jeremiah 17:7-8 NIV)

What does this mean for us in this new week? It simply means when we face an avalanche of unforeseen troubles and stressful situations, we have made a conscious decision to be planted by the river of living water. We are reading and meditating on the Word of God daily, staying hydrated by the river of Living Water, placing our trust and confidence in The Master Gardener, and placing our roots deep so we can withstand the heat (the pressures and struggles of each day). Even in the midst of the heat, we are assured The Master Gardener will allow us to continue to produce fruit, in fact, much fruit to the glory of God.

Stay hydrated this week by drinking long and deep from the Living Water! 🪴🌻☕☕

Sunday Snippet Reflections (SSR)

1. How did today's Snippet speak into my "soul" garden?

____
____
____
____

2. Application to my "soul" garden.

____
____
____

3. One thing I am grateful for:

____
____
____

4. This week, I will commit to pray for:

____
____
____

5. Answered prayers:

____
____
____

Bible/Sermon Notes Date:____________

Scripture:

____

____

Key verses:

____

____

____

Notes:

____

____

____

____

____

____

____

____

____

____

____

____

____

____

Sunday Snippet 36

Do you ever feel like you're just in survival mode? You're just trying to make it to work, through work, and go home, only to know you gotta do it all again tomorrow. 🤣 That's why I'm thankful for Sunday, the one day I don't have to feel I'm in survival mode. I can sip my coffee slowly, eat my breakfast in a relaxed manner, get dressed, and rest in God's presence.

On a side note, are we supposed to be singing, "Oh, the weather outside is frightful? " My poor baby plants! 😩 Honestly, we knew spring-like temperatures were never real. I believe it's just a way God brings us out of our winter slumber. 🧑‍🌾 He gets us thinking about gardening, ordering seeds, cleaning up, and preparing everything for harvest. Yet, we know in our hearts, spring never sprang, and all our precious plants must come inside when temperatures change. The warm weather stirred within us a happy feeling. It motivated us to "wake up" from our winter hibernation. For a moment, we felt invigorated under the sun. 🌞 Now it's cold, and there's a mad dash to cover plants that believed spring had sprung and bring in the baby ones. One day soon, spring will really be here! Our plants will thrive because the conditions will be right, and then comes the harvest. So, hang on, friend. Keep stoking the fire inside your gardening heart.

Speaking of stoking the fire, it's hard sometimes to

stoke the fire of our spiritual garden, isn't it? Oftentimes, we feel the drowning rain of circumstances beating down on the embers that are barely flickering and we wonder how we can get those embers to flame up once again.

The Master Gardener says:

> "Dear brothers and sisters, when troubles of any kind come your way, consider it an opportunity for great joy. For you know that when your faith is tested, your endurance has a chance to grow. So let it grow, for when your endurance is fully developed, you will be perfect and complete, needing nothing." ... (James 1:2-4 NLT)

My dear friend, keep stoking your spiritual fire. Keep persevering! Cold snaps only make us stronger, especially when The Master Gardener is in charge of our "soul" garden. He knows EXACTLY where to place us, and He always covers us with His love, patience, and care. Isn't that something!?!

This week, let's focus on stoking our spiritual fire, even amidst an unexpected cold snap. Let's keep feeding our flickering embers those things that will allow the tiniest of flames to burst into a glorious, glowing fire, bringing The Master Gardener honor and glory.

Keep your fire burning, my friend!

Sunday Snippet Reflections (SSR)

1. How did today's Snippet speak into my "soul" garden?

____

____

____

____

2. Application to my "soul" garden.

____

____

____

3. One thing I am grateful for:

____

____

____

4. This week, I will commit to pray for:

____

____

____

5. Answered prayers:

____

____

____

Bible/Sermon Notes Date:___________

Scripture:

____

____

Key verses:

____

____

____

Notes:

____

____

____

____

____

____

____

____

____

____

____

____

____

____

Sunday Snippet 37

It's Sunday. Did you stir the ashes of your soul this week and kindle a flame? You know, sometimes when you stir the ashes, you only get smoke, but it's a start because that means your soul is still hot. Keep stirring, and soon the flames will roar! 🙂

Spring weather is always changing. We've experienced wind and rain sprinkled with sunshine all in the same week. My plants have been covered, uncovered, and brought inside the house. I've had to constantly change their environment for them to continue to thrive. As the keeper of the garden, I'm doing what is best for my seedlings. Moving them may help them produce fruit. It's up to me, the keeper of the garden, to decide how to help my seedlings develop. Of course, seedlings don't whine when I make these decisions. They just continue to grow. 🪴🌻

Funny how when The Master Gardener changes the environment of our "soul" garden, we truly don't see it working for our good. Maybe the environment The Master Gardener is taking you to isn't what you expected. Health changes, job change, added bills, taking care of aging parents, stress overload, and the list goes on. Your conditions have changed significantly, and you wonder if you'll survive and thrive.

The Master Gardener knows what is best! Often, we

get too comfortable in our current environment, and The Master Gardener moves us to regenerate growth. Sometimes, our tears are what water our "soul" garden, allowing us to find joy in our new environment.

But what if The Master Gardener just moves you into a new environment to test your ability to withstand hardship? To see if you trust that His decision to move you is really what is best. I'm experiencing a new environment right now. I've decided I want to be just like my seedlings. I want to survive and thrive in whatever environment The Master Gardener places me.

We are never placed into a new environment without The Master Gardener providing guidance from the Manual. Here's what I know. Zephaniah 3:17 ESV carries encouragement. It reminds me He is "in the midst" to strengthen me, support me, protect me, defend me, and deliver me.

> "The LORD your God is in your midst,
> a mighty one who will save;
> he will rejoice over you with gladness;
> he will quiet you by his love;
> he will exult over you with loud singing." ... (Zephaniah 3:17 ESV)

My dear friend, whenever The Master Gardener moves us into a new environment, it's really for our good and ultimately His glory. Let's decide to survive and thrive wherever The Master Gardener places us! 🪴🌻☕☕

Keep stoking the fire!

Sunday Snippet Reflections (SSR)

1. How did today's Snippet speak into my "soul" garden?

____
____
____
____

2. Application to my "soul" garden.

____
____
____

3. One thing I am grateful for:

____
____
____

4. This week, I will commit to pray for:

____
____
____

5. Answered prayers:

____
____
____

Bible/Sermon Notes Date:___________

Scripture:

____

____

Key verses:

____

____

____

Notes:

____

____

____

____

____

____

____

____

____

____

____

____

____

____

Sunday Snippet 38

🌧️🌞❄️ It's a typical week in NC, keeping us on our gardening toes! I hope you have experienced some extra sunshine in your life this past week. Spring is coming and I can't wait!

How are your plants? Mine are coming along. I'm sure they would like more sun and less cool weather. Hopefully soon. I know I need more sunshine. I'm not a fan of too many rainy days in a row.

Have you ever looked closely at a tiny seedling? Have you ever wondered, "How in the world is this tiny plant standing?" It stands straight and strong. The tiny stem is holding out its leaves as if welcoming the gardener to see its beauty and strength. As the gardener, we provide water and nutrients to help our plants grow bigger and stronger. Yet it's not what we see with the naked eye that's really holding the plant in place. It's deeper.

Bet your first thought was, "It's the roots!" Yep, it's the root that allows the seedling to stand strong. You know, if a plant has a weak root system, it usually doesn't make it when winds howl and stormy weather prevails. Roots matter!

I've been thinking about the roots in my "soul" garden and how I need to grow them deeper in my relationship with The Master Gardener. Roots matter! The storms in our lives vary from one season to another, yet each storm tests our root system.

The Master Gardener places emphasis on where our roots are to be and even gives us a reason why. Deep roots (faith roots) produce an abundance of thankfulness. We are told in Colossians 2:7 NLT to "Let your roots grow down into him, and let your lives be built on him. Then your faith will grow strong in the truth you were taught, and you will overflow with thankfulness."

Let's focus on what soil we are placing the roots of our "soul" garden in this new week. Let's choose to grow deeper in our relationship with The Master Gardener. Roots, shoots, and thankfulness! 🪴🌻🤠☕☕

Remember...roots matter!

Sunday Snippet Reflections (SSR)

1. How did today's Snippet speak into my "soul" garden?

2. Application to my "soul" garden.

3. One thing I am grateful for:

4. This week, I will commit to pray for:

5. Answered prayers:

Bible/Sermon Notes Date:___________

Scripture:

____

____

Key verses:

____

____

____

Notes:

____

____

____

____

____

____

____

____

____

____

____

____

____

____

Sunday Snippet 39

If you're anything like me, you've enjoyed the sunshine this past week. Its rays seemed to massage my heart, chasing away burdens, awakening my "soul garden" to spring, and the possibility that it is really here. 🌞🌻

My seedlings are soaking in the sunshine too. They are getting stronger every day. So, I have a crazy question. Do you ever talk to your plants? I do! They are great listeners, 🤣 and it's when I'm chatting to them that I see any issue that particular plant might be displaying. I'm sure they don't know my voice or really care if I speak, but for some odd reason I find pleasure in talking to them.

Wondering what I say? 🫢 I say such things as: "My, you've grown." Or "You look thirsty...need some water?" And lately, "Doesn't the sunshine feel amazing!" Not once have any of my plants said, "You're the greatest gardener we know!" Or "I was about to thirst to death! Thank you for the water." Although I know my plants will never talk, it doesn't deter me from talking to them. 🤣🌻🌻

I want you to get your Manual. Turn to John 20 and read the whole chapter. I'm pulling out my favorite verses starting in verse 14.

> She turned to leave and saw someone standing there. It was Jesus, but she didn't recognize him. "Dear woman, why are you crying?" Jesus asked her. "Who are you looking for?"

> She thought he was the gardener. "Sir," she said, "if you have taken him away, tell me where you have put him, and I will go and get him."
>
> "Mary!" Jesus said.
>
> She turned to him and cried out, "Rabboni!" (which is Hebrew for "Teacher"). ... (John 20: 14-16 NLT)

Mary Magdalene was distraught. She believed someone had taken her Lord away, stolen His body. I'm sure she was weeping, and through blurred vision, saw a man who she assumed was the gardener. Get ready! Here comes my favorite verse, verse 16. The Master Gardener calls Mary's name. Immediately, she recognized the voice. His voice. Her Lord's voice. Her Shepherd's voice. Her Savior's voice. The voice she loved. He called her name. He knew her name.

He wasn't just the gardener. He was and is The Master Gardener. You know, if Mary hadn't had a relationship with Jesus, she would have never recognized His voice. My friend, The Master Gardener, knows your name, too. He knows mine. He's gently calling your name. He's telling you how much He loves you and how He will never leave you or forsake you.

Just stop and get still. Do you hear Him? He knows YOUR name. He knows mine. He stands at the gate of your "soul" garden, gently calling your name, desiring to talk, to encourage, to heal, and to save. Isn't our Master Gardener wonderful?

Of all the names He knows, He knows yours. He knows mine. Isn't it comforting to know He loves you so much that He desires a relationship with you? He's not just the Gardener, He's The Master Gardener! Hallelujah, my friend! Hallelujah!!

✝️☕☕

Sunday Snippet Reflections (SSR)

1. How did today's Snippet speak into my "soul" garden?

2. Application to my "soul" garden.

3. One thing I am grateful for:

4. This week, I will commit to pray for:

5. Answered prayers:

Bible/Sermon Notes Date:___________

Scripture:

___

___

Key verses:

___

___

___

Notes:

___

___

___

___

___

___

___

___

___

___

___

___

___

___

Sunday Snippet 40

🌻 It's Sunday! The sun will be shining, which causes me to feel so much joy! 🌞 Isn't it funny how a sunny day can literally chase the blues away? Think about it. Weather affects our emotions, sometimes it causes physical changes (aches, pains, allergies), and it often dictates how we plan our day. The weather can be unpredictable! Without warning, storms can blow up rather quickly on a steamy summer day. Hail damages, winds destroy, water floods, drought stunts growth, and extreme cold causes many issues. Weather can be stressful, yet we rarely think about how it affects our lives, unless we are a plant. 🤣

Did you know plants feel stress? Weather affects every aspect of a plant's life. Plant stress has been defined by Hartmut K. Lichtenthaler (1996) as "any unfavorable condition or substance that affects or blocks a plant's metabolism, growth, or development." Plants feel stressed out on occasion. Do you?

What is the one thing (okay, the multitude of things) causing you stress right now? Get a pen and some paper. Write them down. I'll wait. If you're like me, when you saw the list, you felt stressed. 🤣😩

As a gardener, you are constantly checking your plants for signs of growth or any environmental stressors that could harm the plant's future. I want my plants to thrive! So, as the gardener, I look for ways to reduce the stress on my plants.

The Master Gardener does the same for us. Look at the list you just wrote down (seriously, you need to write it down). The list can be long because, honestly, we can just feel downright overwhelmed by life. If we are brutally honest, we would believe The Master Gardener has given up on our "soul" garden and moved on to another garden more worthwhile, one with more growth, one with more commitment, one with more potential for a larger harvest. Maybe you feel The Master Gardener is tired of trying to keep your "soul" garden weeded and cared for and has moved on. Why would The Master Gardener keep weeding out sin and keep watering you when there's a drought in your life, fertilizing your "soul" garden with words from the Manual, and pruning you to cause growth and enhance fruit production? Why? Because The Master Gardener loves you. Hard to believe someone could love us the way The Master Gardener does, isn't it? There's not one stressful situation that The Master Gardener isn't aware of in your life. There's not one He doesn't understand or can help you find a way out of. Why? Because He loves you! He loves me! Look at what the Manual says:

> And I am convinced that nothing can ever separate us from God's love. Neither death nor life, neither angels nor demons, neither our fears for today nor our worries about tomorrow—not even the powers of hell can separate us from

> God's love. No power in the sky above or in the earth below—indeed, nothing in all creation will ever be able to separate us from the love of God that is revealed in Christ Jesus our Lord. ...
> (Romans 8:38-39 NLT)

Let the sun shine in your "soul" garden today because The Master Gardener is still working and caring for you, and oh, how He loves you and me! Stressed out, pooped, tired, and weary. His love endures forever! 🧑‍🌾☕🌻☕

Sunday Snippet Reflections (SSR)

1. How did today's Snippet speak into my "soul" garden?

____
____
____
____

2. Application to my "soul" garden.

____
____
____

3. One thing I am grateful for:

____
____
____

4. This week, I will commit to pray for:

____
____
____

5. Answered prayers:

____
____
____

Bible/Sermon Notes Date:____________

Scripture:

____

____

Key verses:

____

____

____

Notes:

____

____

____

____

____

____

____

____

____

____

____

____

____

____

Sunday Snippet 41

Hope all is well with you and yours! What glorious weather we had last week! Writing down the things that stressed me out really helped me see I was wasting a lot of time on trivial things. What about you? I understand now how much attention I was giving things that, in light of eternity, did not matter. It was freeing for me to allow The Master Gardener in the gate of my "soul" garden. He helped pull those nasty weeds of the things that stressed me out of my 'soul garden. 'Believe me, the roots were deep on some!

Walk with me through my garden this morning. Isn't it peaceful? Hear the birds singing? They seem to ALWAYS have a song in their heart. Never fearful of praising their Creator. The air is crisp this morning, but the plants don't seem to mind. Take a gander at my plants. They are all working on growing. Not all of them are the same size, but you see their effort. Isn't this just so relaxing? I love my garden! 🪴🪴

Come sit with me. Got your coffee? ☕ I have a question for you. What do you really care about? Now write it down. Yes, really write it down. I care about my family, my friends, my animals, my relationship with The Master Gardener, my neighbors, my students, and my plants. Those were the first things that popped into my mind that I quickly wrote down.

Look at your list. Now, suppose one or more things on your list do not perform like you want. What's your plan? Do you get rid of it? Or do you move in closer to see what the problem is?

Remember our tour through my garden? I showed you my plants, and some of my sugar peas are super tall, and some, on that same row, appear to be stunted in growth. 😬 As a gardener, I realize plants take time to grow. Different things affect their growth (where they are planted in relation to the sun or if they get too much shade), but as long as they are alive and growing, I don't just yank them up out of the ground and discard them.

Look carefully at your list of things you care about. Read each one out loud so you can hear it. Ever felt like discarding something on your list? 🤣 We all have, because we're human. But if the things you care about are still growing, still striving to become something (like my plants), still loving you regardless of your actions or reactions, then don't yank them from your "soul" garden.

The Manual says in 1 Peter 4:8 NLT, "Most important of all, continue to show deep love for each other, for love covers a multitude of sins."

Love isn't an emotion (although it can be part of it). Love is an action. It's an attitude about life, about The Master Gardener, and all the things we put on our list. Love is what gives us a peace within the gates of our "soul" garden. Love is why The Master Gardener came to your garden and mine.

Thought for this new week: Don't discard what is still trying to grow. Instead, water it with love. 🌻☕☕

Sunday Snippet Reflections (SSR)

1. How did today's Snippet speak into my "soul" garden?

____

____

____

____

2. Application to my "soul" garden.

____

____

____

3. One thing I am grateful for:

____

____

____

4. This week, I will commit to pray for:

____

____

____

5. Answered prayers:

____

____

____

Bible/Sermon Notes Date:____________

Scripture:

Key verses:

Notes:

Sunday Snippet 42

It's Sunday. It's a good day to pull back, sit on the porch, and talk. 🤠 The weather is crazy again. Heat. Storms. Chilly. My plants are confused.

Today, I invite you to come sit on the porch with me. We haven't talked on the porch in a long time. Grab your coffee. ☕☕

I've been thinking a lot about gardening. There are so many ups and downs. You plant a host of seeds, and only half germinate. You water, fertilize, and care for each tender plant, yet some die. It's frustrating. Yet, as gardeners, we move forward, taking the good with the bad. Thankful for the seeds that germinated. Grateful for the harvest to come.

Life is often like our gardens. We plant hopes and dreams and invest our time and energy in them, only to see some die. It's heartbreaking!

However, as soul gardeners, we are always grateful for the dreams that germinated and are growing. We take great care to water, fertilize, and prune these blessings. Why? Because for every broken dream, a new "dream" seed is formed and planted into our hearts.

Brokenness and blessings. Strange combination, isn't it? I have often wondered why The Master Gardener chose brokenness and blessing to work together. How can these seeds be planted in the same pot and produce something of worth?

Seeds that do not die do not grow. That's a gardening fact. Because of our brokenness, we are blessed. Let that sink in. Brokenness reveals beauty. It allows The Master Gardener to come close and to inspire us to rise up from the ashes of brokenness.

The Manual says this about brokenness and blessing in Psalm 34:18 NLT. "The Lord is close to the brokenhearted; he rescues those whose spirits are crushed...." and that:

> ...he will give a crown of beauty for ashes,
> a joyous blessing instead of mourning,
> festive praise instead of despair.
> In their righteousness, they will be like great oaks
> that the LORD has planted for his own glory. ...
> (Isaiah 61:3 NLT)

Do you see it? From these two sections in the Manual, The Master Gardener gently whispers your name and says that even from the seed of brokenness, He will cause it to rise up and be as strong as a great oak, and that great oak will give The Master Gardener honor and glory.

Some days we feel more broken than others, but remember, The Master Gardener knows about brokenness. He comes near, not with words of condemnation, but with words of hope and assurance that beauty is on the horizon. So, just hold on. Just hold on. 🌻☕☕

Sunday Snippet Reflections (SSR)

1. How did today's Snippet speak into my "soul" garden?

____

____

____

____

2. Application to my "soul" garden.

____

____

____

3. One thing I am grateful for:

____

____

____

4. This week, I will commit to pray for:

____

____

____

5. Answered prayers:

____

____

____

Bible/Sermon Notes Date:___________

Scripture:

Key verses:

Notes:

Sunday Snippet 43

It was a busy week! And I survived! 🤣 Have you ever noticed that busyness often keeps us away from important things? We can become SO busy that we forget things of importance, especially in the garden.

How's your garden? Mine is almost completely planted, but the weeds desire to take over. Busyness has led to neglect, and in turn has made my garden work harder.

It won't be long until the garden pests will roll out their red carpet and make their grand entrance into my rows of plants. So, what does a gardener do? We start spraying something to take those pests out! But is that always necessary? Harsh chemicals are stressful to your plants, and we know that stress causes unhealthy plants, which do not thrive. I'm researching a different approach to pest control this year, and it starts with amending my soil. Pest control and soil amendment go hand in hand. I'll keep you posted on how this works. 🤠

In my little bit of research, I've come to understand one thing: you cannot be an absent gardener and expect a harvest. You have to be present for good things to happen and to handle unwanted pests (and realize not all bugs are bad). Gardening is hard work! It's not for the faint of heart! It's a commitment if you want a harvest.

My physical garden is a lot like my "soul" garden because they both require a commitment to receive a harvest. We must amend the soil in our "soul" garden daily! The pests of doubt, fear, anger, hopelessness, discouragement, anxiety, loneliness (and a host of others) need a daily spraying of one important chemical. I found this solution in the Manual. Galatians 5:22-23 NIV reads, "But the fruit of the Spirit is love, joy, peace, forbearance, kindness, goodness, faithfulness, 23 gentleness and self-control.)

Powerful chemicals! These chemicals for our "soul" garden can be used individually or mixed together. There are two other ingredients that should always be mixed into these: prayer and thankfulness. Use these liberally because they work together with the others to destroy the pests that show up in our "soul" garden.

Soul amending. It's not easy to love or forgive. It's not easy to encourage or sow seeds of kindness. That's why we need to be amending our "soul" garden daily by reading the Manual. Reading and following the Manual will help destroy unwanted pests, and in turn, the soil in our "soul" garden becomes richer so fewer pests want to enter the rows we've planted with love, joy, and peace.

I've heard the best medicine for a garden is the Gardener! But I know the best medicine for our "soul" garden is The Master Gardener.

Plant seeds. Pull weeds. Practice good deeds. 🪴🌻☕☕

Enjoy the day God has given you! ☕☕

Sunday Snippet Reflections (SSR)

1. How did today's Snippet speak into my "soul" garden?

2. Application to my "soul" garden.

3. One thing I am grateful for:

4. This week, I will commit to pray for:

5. Answered prayers:

Bible/Sermon Notes Date:___________

Scripture:

Key verses:

Notes:

Sunday Snippet 44

Have you stopped and thanked God for providing another day for you? Have you stopped to listen to His voice? Amazing how we forget our Creator when life shakes us to the core, when faced with uncertainty and hurt, when we place ungodly things above Godly, and before long God is left behind while we seek other means of keeping our "soul" garden alive.

I have worked countless hours in my garden! Planting, watering, looking for destructive insects, and pulling weeds. And yet, it never fails, one or two plants die while others in the same row thrive. How is that? I planted them and took care of them. I provided everything they needed to bear fruit, and yet, they didn't make it. And it's always a favorite plant that decides to pull out of the race. 🤣

Funny how we often only see the negative things in life, isn't it? I have over a hundred plants that are thriving, yet my focus is on the few plants that died. Isn't this so true in our "soul" garden? We focus on the few negative events and forget all the good things God has done. Why is that? The world provides no medicinal cures, but we rush to it when the winds of negativity blow through the rows of our "soul" garden, crushing a few feelings here and there. When we focus on the loss and the negativity, we completely lose sight of how The Master Gardener is still tending and caring for our "soul" garden.

The Master Gardener cares about you and me! He cares when the winds of weariness howl, when hurtful words find their way to a row of already raw emotions, and He cares immensely when a drought strikes our "soul" garden. Open the gate to your "soul" garden and take a look around. How's it doing? Are you experiencing weeds of worry? An unforeseen drought? Are you seeing rows of weariness? Have you noticed a few "dead" plants squashed by hurtful words?

Guess what? A few dead plants don't mean failure in a garden, and the same concept is relevant in our "soul" garden. A few bad moments don't mean a failed life. There's truly more good than bad if we look for it.

When you feel overwhelmed in tending to your "soul" garden, remember the Manual gives hope. Here's what I found in Psalm 107:35 NLT, "But he also turns deserts into pools of water, the dry land into springs of water."

The Master Gardener is always ready to turn on the sprinklers of blessings to restore growth and hope. So, soak it up! Thrive wherever The Master Gardener has planted you!

My friend, it's a new day! A new week filled with new opportunities. Allow The Master Gardener to turn the sprinklers of blessings and thankfulness on, and when He does, you will begin to feel refreshed and restored in the garden of your soul. 🌱🌻☕☕

Sunday Snippet Reflections (SSR)

1. How did today's Snippet speak into my "soul" garden?

2. Application to my "soul" garden.

3. One thing I am grateful for:

4. This week, I will commit to pray for:

5. Answered prayers:

Bible/Sermon Notes Date:___________

Scripture:

____

____

Key verses:

____

____

____

Notes:

____

____

____

____

____

____

____

____

____

____

____

____

____

____

Sunday Snippet 45

Guess where I am? No, not the beach or the mountains. I'm sitting in my greenhouse. It's a wonderful place! So many hours were spent waiting for this build. Now I can spend hours in here planting seeds while waiting for new life to appear.

Waiting for seedlings to appear is hard! We plant seeds on Saturday and check on Tuesday to see if we can locate a tiny sprout. 🤣 We're waiting yet watching for any sign of growth. Any gardener knows one thing for sure: not every seed produces life. Some seeds never surface past the hole you put them in, while other seeds do their job. They die first to produce a harvest, while we watch and wait.

Isn't life also like a waiting room? We wait for so many things! We wait for a spouse, a home, children, a job, a doctor's report, an answer to prayer, and of course, retirement. The waiting room of life is really our best teacher. We all know full well that every single thing we are waiting for will come to fruition. But still we wait. Wait and watch, watch and wait. Wearisome, isn't it?

Well, this morning when I got into the greenhouse, I found The Master Gardener was already here. Waiting. 🤠 He was waiting on me to join Him, to converse with Him, and to listen to His counsel. It's been a busy year, a busy month, and the last few days have been filled to the brim with busyness. I'm glad The Master Gardener visited with me today. I'm glad He waited for me to come, as weary as I am. He watched and waited.

When I opened the greenhouse door, He smiled. I knew He was glad to see me. I told Him all the things on my mind, and He assured me that the Manual had a solution. Wow! When The Master Gardener reveals and speaks His truth in your "soul" garden, it's comforting and healing.

He told me it was okay to feel weary and that it wouldn't last forever. He spoke:

> "But those who wait for the Lord [who expect,
> look for, and hope in Him]
> Will gain new strength and renew their power
> They will lift up their wings [and rise up close to
> God] like eagles [rising toward the sun];
> They will run and not become weary,
> They will walk and not grow tired." ... (Isaiah 40:31
> AMP)

The waiting room of life often causes weariness, but while waiting, The Master Gardener renews our hope, our strength, and our commitment to keep running the race. 🪴🌻

Waiting rooms are really the best teachers. Blessings on your new week! ☕☕

Sunday Snippet Reflections (SSR)

1. How did today's Snippet speak into my "soul" garden?

___

___

___

___

2. Application to my "soul" garden.

___

___

___

3. One thing I am grateful for:

___

___

___

4. This week, I will commit to pray for:

___

___

___

5. Answered prayers:

___

___

___

Bible/Sermon Notes Date:___________

Scripture:

____

____

Key verses:

____

____

____

Notes:

____

____

____

____

____

____

____

____

____

____

____

____

____

____

Sunday Snippet 46

Well, do you feel renewed? Or are you still a little weary? I think I'm still a little weary. As the old saying goes, "You don't know how tired you are until you quit." You work hard all day nonstop, and the second you sit down, you know you have zero plans of getting back up.🤣 That's where I'm at. We all need rest!

Gardening can often feel like a one-man show. It can be hard to enlist others to help joyfully if they don't share the same passion. Of course, they don't mind sharing in the harvest. Eating that juicy tomato sandwich, loving those freshly cooked green beans, reaching out for every single piece of crunchy fried squash you lay in the tray. You know the ones. 🤣🤣 and the funny thing is, we don't mind. In fact, it makes us feel good because we planned, planted, watered, and The Master Gardener provided the increase. All the hard work paid off!

We often feel like life is a one-man show. We sleep. We get up. We go to work. We come home and prepare to do it all again the next day. Read those sentences over and over, and you feel exhausted. Life, your life and mine, was never meant to be a one-man show. We were meant to work hand in hand with The Master Gardener. The Manual says,

> Then Jesus said, "Come to me, all of you who are weary and carry heavy burdens, and I will give you rest. Take my yoke upon you. Let me teach you, because I am humble and gentle at heart, and you will find rest for your souls. For my yoke is easy to bear, and the burden I give you is light." ... (Matthew 11:28-30 NLT)

I've never plowed with oxen. What I know is this: a farmer never puts two untrained oxen in a yoke together expecting a plowed field. He puts one seasoned ox with the trainee. Naturally, the experienced ox bears the majority of the load. Since the older one leads, the younger ox does not have to wonder what to do. He learns from his mentor and gains the knowledge and skill to teach others.

This very much resembles discipleship. Isn't it just like The Master Gardener to help us see when we are working alongside Him, the burdens we carry aren't as heavy, and life isn't a one-man show? He's there to teach us, to take the load off, and to renew our spirit. First, we have to be willing to put on the yoke and learn from The Master. That's the hardest part!

This week, let's share the yoke with the One who can train us to become more like Him. 🌻 He makes life easier! ☕☕

Sunday Snippet Reflections (SSR)

1. How did today's Snippet speak into my "soul" garden?

____
____
____
____

2. Application to my "soul" garden.

____
____
____

3. One thing I am grateful for:

____
____
____

4. This week, I will commit to pray for:

____
____
____

5. Answered prayers:

____
____
____

Bible/Sermon Notes Date:___________

Scripture:

____

____

Key verses:

____

____

____

Notes:

____

____

____

____

____

____

____

____

____

____

____

____

____

____

Sunday Snippet 47

Feeling better? I am! I feel like I've made some much-needed progress in my life. You know, life can get messy on occasion. Sometimes you feel like your life looks like dominoes falling. Once one thing happens, everything else starts to fall. I'm currently setting my life's dominoes back up. You are too! I see you working on it! 🤠

I know one day something will knock them over again, but for now, I'm busy reconfiguring them with a fresh set of eyes. Reconfiguring those life dominoes on solid ground and setting them up strategically.

Gardening is good. It's a full-time job! You have to be strategic in how you plan, water, weed, and harvest. It's all about timing. Timing is crucial for gardeners. Plant too early and you put yourself in danger of frost, potentially losing your crop. Plant too late and you may not have a harvest. We even weed strategically! Daily pulling weeds, watering when the heavens aren't dropping those much-needed droplets, and looking intently for a harvest.

You know who else is strategically planning to do a little digging in your "soul" garden? The Enemy! He sees you and me getting our life dominoes set back up. He knows the last domino crash caused you and me some serious grief, but he also knows we became stronger through it, and now we are setting those life dominoes on solid ground. Yep! We learned a thing or two. 🤣🤣

The Master Gardener is on the move strategically, too. He's provided us with some "soul" garden armor. Armor that protects us from the strategies of the Enemy. Isn't that just like The Master Gardener? Always looking out for the gardener. 👩‍🌾 Look in the Manual at Ephesians 6:10-11 NLT, "A final word: Be strong in the

Lord and in his mighty power. 11. Put on all of God's armor so that you will be able to stand firm against all strategies of the devil."

The Master Gardener keeps us safe from the one who is constantly trying to push down that first domino. We have to be actively "putting on" the armor every day! The Master Gardener has always had a strategic plan for us. It's in the Manual. In Jeremiah 29:11 AMP it reads, 'For I know the plans *and* thoughts that I have for you,' says the Lord, 'plans for peace *and* well-being and not for disaster, to give you a future and a hope.'

Be alert as you reconfigure your life dominoes. Strategically place them within the plan The Master Gardener is showing you, and suit up, friend! The Enemy stands ready for attack, but we'll be wearing armor from The Master Gardener's greenhouse. 🪴🌻

Be strategic this week! ☕☕

Sunday Snippet Reflections (SSR)

1. How did today's Snippet speak into my "soul" garden?

____
____
____
____

2. Application to my "soul" garden.

____
____
____

3. One thing I am grateful for:

____
____
____

4. This week, I will commit to pray for:

____
____
____

5. Answered prayers:

____
____
____

Bible/Sermon Notes Date:__________

Scripture:

____

____

Key verses:

____

____

____

Notes:

____

____

____

____

____

____

____

____

____

____

____

____

____

____

Sunday Snippet 48

Well, He did it again! The Master Gardener brought us safely through another week and blessed us with a brand-new day. It's been a busy week! I've been canning, picking beans, and planting seeds for a "round two" harvest. Oh, and were we not super blessed when The Master Gardener sent rain?! 🌧️🪴

While I was planting more rows of green beans, I said to myself, "Self, what if I only planted a couple of seeds?" Maybe both seeds would grow, but the harvest would certainly be limited. Worst-case scenario, neither seed would sprout. 😩 So, after having that conversation with myself, I planted two rows. 🤣🪴🪴🪴

In life, it's important to plant seeds too. More than a couple! Many months ago, we talked about seeds and seed packets, but sowing seeds of gratitude, I believe, is one of the hardest seeds to sow and grow. Why? Gratitude seeds do not sprout easily. The weeds of discontent and negativity pop up quickly! Instead of pulling those weeds asap, we allow them to grow, sometimes taking over our whole gratitude row, often wiping it out completely! 😥 Funny how we know these weeds are intruders of the good things planted in our "soul" garden, but we get overwhelmed with the task of pulling those pesky things up, and then the unthinkable happens. We succumb to discontent and negativity, and sadly, we stop being thankful. It's easy for this to happen because keeping weeds out of our "soul" garden is hard work!

Thankfully, we have the Manual to give us guidance on how to rid our "soul" garden of such weeds. Hebrews 13:5 NKJV says, "Let *your* conduct *be* without covetousness; *be* content with such things as you have. For He Himself has said, "I will never leave you nor forsake you."

You want your gratitude seeds to sprout? Water them! Every day, water your row with one thing you are thankful for and write it down. Guess what? This will start the process of eliminating the weeds of discontent or negativity. 1 Thessalonians 5:18 NASB says, "in everything give thanks; for this is the will of God for you in Christ Jesus."

Challenge for the week. Write down one thing you are thankful for each day. It's life-changing! We have to be diligent to pull those nasty weeds out of our "soul" garden! I want to have a harvest of gratitude! How about you? So, let's get our seeds of gratitude out and start planting! ☕☕

Sunday Snippet Reflections (SSR)

1. How did today's Snippet speak into my "soul" garden?

2. Application to my "soul" garden.

3. One thing I am grateful for:

4. This week, I will commit to pray for:

5. Answered prayers:

Bible/Sermon Notes Date:___________

Scripture:

____

____

Key verses:

____

____

____

Notes:

____

____

____

____

____

____

____

____

____

____

____

____

____

____

Sunday Snippet 49

Chaotic times we are living in. 🤭 I'm not much into chaos. I'm happy working in my garden, picking beans, canning salsa, pickling radishes, making jelly, and pulling a weed here and there. 🧑‍🌾 The garden, although hard work, is where I have found peace and contentment. What about you? Where's that quiet place that brings rest to your soul?

I looked up chaos in the dictionary. It means complete disorder and confusion. As I walk through my garden this morning, everything inside this gate is safe. With the exception of a few bugs snacking, 🤣 there are no predators. It's not chaotic in my garden, it's peaceful.

Can gardens have chaotic times? Absolutely! If I don't manage the garden, it becomes a mess! Weeds, bugs, lack of water, veggies rotting on the vine, and plants die. It's because they are not cared for. So yes, a garden can experience chaotic times.

While I was assessing the garden chores and walking to the garden this morning, guess who was there waiting for me at the gate? Yep! The Master Gardener. Oh, it was delightful seeing Him. His smile was warming, and He complimented my efforts in keeping my garden in order. I told Him I couldn't have done it without Him. 😂 🥒🫑🌶️ We both chuckled. Meeting The Master Gardener each day at the gate fills me with peace. He walked beside me as I watered my plants. We talked and talked. Do you ever just talk to The Master Gardener? He is the keeper of our "soul" garden, you know. He's interested in every detail of our lives!

As we talked, I told Him about the prices of necessary items exceeding what I could afford. He nodded. I told Him how I felt about the world and how, at times, I was feeling overwhelmed. He nodded.

I talked and talked. He listened, nodded, and laughed. We had a wonderful time! As I was finishing up, The Master Gardener took my hand and looked at me. His eyes were full of love and compassion because He alone knew my heart. He had listened to me intently, and now I needed to hear Him.

This is what He said to me. In the Manual, I would find these words written. They were meant for me today. He told me to listen to Him carefully, and I did! 2 Timothy 1:7 AMP says, "For God did not give us a spirit of timidity *or* cowardice *or* fear, but [He has given us a spirit] of power and of love and of sound judgment *and* personal discipline [abilities that result in a calm, well-balanced mind and self-control]."

He told me to highlight it in my Manual, to memorize it, and to hide it in my heart. As I finished up my chores, He assured me that He was always with me. In fact, He said He never leaves my "soul" garden. He's always there working, keeping chaos at bay, holding me together when I'm falling apart, watering my soul with His words, and holding my hand with His nail-scarred hand to assure me He's more than worthy of all the praise because He is The Master Gardener.

Dear friend, I pray you meet The Master Gardener at the gate of your "soul" garden this morning. Tell Him everything on your mind, but after that, lean into Him, be still, and listen. He's got a word for you!

Sunday Snippet Reflections (SSR)

1. How did today's Snippet speak into my "soul" garden?

2. Application to my "soul" garden.

3. One thing I am grateful for:

4. This week, I will commit to pray for:

5. Answered prayers:

Bible/Sermon Notes Date:___________

Scripture:

____

____

Key verses:

____

____

____

Notes:

____

____

____

____

____

____

____

____

____

____

____

____

____

____

Sunday Snippet 50

Good morning! How's every little thing? I know you have been busy doing life. Me too. Did you make time to meet The Master Gardener each day? Don't you find it amazing and beautiful how He waits at the gate of our "soul" garden each day? And the smile He gives when He sees us? There's not a word to describe it. 🤠

I've certainly been busy in my garden. Picking beans, squash, cucumbers, and peppers, and then preserving them. Harvest time for the gardener is a special time. It's kinda like receiving the gold medal in the Olympics. 🏅 🏅 Of course, there are no spectators, no one cheering you on as you pick beans or labor preserving garden goods, no coaches in-between the rows to hand you water and encourage your weary soul, and no one from the Garden of America Club is there to place the medal around your neck as you stand proudly with your bucket of beans, cucumbers, and peppers. Nope. It's just you feeling awesome and blessed because of your perseverance.

When I planted my garden, it required lots of labor on my part. I tilled the ground, made the rows, and planted the seeds. If no rain occurred, 🌧 I got out there and watered. 👩‍🌾 I did my part, but I couldn't make my seeds grow. They had to choose to die so they could live. Strange how The Master Gardener says the same thing about you and me. John 12:24 NLT says, "I tell you the truth, unless a kernel of wheat is planted in the soil and dies, it remains alone. But its death will produce many new kernels—a plentiful harvest of new lives."

Sunday Snippet Reflections (SSR)

1. How did today's Snippet speak into my "soul" garden?

____

____

____

____

2. Application to my "soul" garden.

____

____

____

3. One thing I am grateful for:

____

____

____

4. This week, I will commit to pray for:

____

____

____

5. Answered prayers:

____

____

____

Bible/Sermon Notes Date:___________

Scripture:

Key verses:

Notes:

Sunday Snippet 51

Did you hear those "soul" garden cheerleaders this week? Gardening is hard work! The heat. The bugs. Rain or the lack thereof, but then the harvest. Harvest time is the best!

The Master Gardener is standing in my garden. He's smiling, knowing how hard I have worked, but He's smiling for another reason. The Master Gardener is smiling because my "soul" garden is in better shape. Weeds are being removed. New growth is happening. How about your "soul" garden? It's hard work for sure! Despite that, The Master Gardener is waiting and willing to come alongside you and me.

The Master Gardener and I walk and talk, stopping along the way and picking some veggies. I find myself, each day, desiring to know Him more to reflect His image. It's hard to explain this desire in my heart.

After a while, The Master Gardener opens the garden gate. As we leave, The Master Gardener opens the Manual. He hands it to me and asks me to read. Jeremiah 24:7 NCV says, "I will make them want to know me, that I am the Lord. They will be my people, and I will be their God, because they will return to me with their whole hearts."

Dear friend, there will be times when we feel lonely, afraid, or discouraged, but in the midst of everything, our deepest desire is to know The Master Gardener more each day. Why? Because He puts that desire within us. Dear friend, He's always there, attentive and trustworthy.

Sunday Snippet Reflections (SSR)

1. How did today's Snippet speak into my "soul" garden?

2. Application to my "soul" garden.

3. One thing I am grateful for:

4. This week, I will commit to pray for:

5. Answered prayers:

Bible/Sermon Notes Date:___________

Scripture:

Key verses:

Notes:

Sunday Snippet 52

🎬 Can you believe it? We've been together in the garden for 52 weeks. Each week, we have stood at the garden gate and talked. 👩‍🌾 I'm glad you met me this morning. I brought my coffee. ☕ I know you have yours. ☕🤣

So, my dear friend, how's your garden doing? I'm picking beans every few days, and cucumbers have a mind of their own. 🥒🥒🥒 They are an everyday picking.

Still waiting on my tomatoes. 🍅🍅 Waiting is hard, but I keep walking through my rows of tomatoes looking for one ripe tomato. When you really love tomatoes, you check their progress daily. As a gardener, you know that sooner or later, a ripe, luscious tomato will be hanging from the vine. Until that day, we look and wait, knowing our reward is coming. 🍅🍅

The Master Gardener is standing at the gate of our "soul" garden. He's there every day, walking through the rows, inspecting the fruit, pulling weeds, and removing anything that would cause harm to potential growth. He's ALWAYS in our "soul" garden waiting. Waiting on us to produce fruit that ultimately will bring honor and glory to Him. Whether we acknowledge Him or not is totally up to us. The Master Gardener will never push Himself on us. He waits patiently for me and you to join Him every day.

One day, The Master Gardener will return with His Father to gather the harvest from His garden. What a day that will be! He's worked in your "soul" garden and mine, relentlessly calling us out to plant better seeds, to rid ourselves of worthless weeds, untangle us from vines that were binding us and hindering growth, and guarded the gate of our "soul" garden day and night so the enemy couldn't destroy the work He was and is doing. WOW!

The Master Gardener will not announce His return when He comes to gather His harvest. His desire is for us, the fruit, to be ready. In 52 weeks, we have worked on our "soul" garden together. We have gained knowledge from the Manual on how to tend our garden, and now we come to harvest time.

Just like me, when I walk through my tomatoes looking for one that is ripe, The Master Gardener is looking for a harvest that is ripe for the picking. I found two important facts in the Manual. He loves us, and He's coming back for us. John 3:16 NLT says, "For this is how God loved the world: He gave his one and only Son, so that everyone who believes in him will not perish but have eternal life." Matthew 24:44 NLT teaches us, "You also must be ready all the time, for the Son of Man will come when least expected."

What a difference a year makes, my friend! The Master Gardener will return for His harvest! Keep searching for Him. Keep listening to His voice. Keep tending to your "soul" garden. And most of all, keep meeting Him at the gate. In fact, push the garden gate open and let The Master Gardener sit on the bench. It's a great thing when you know He resides in your "soul" garden. I've experienced it!

I pray we continue to meet at the garden gate in days and weeks to come. Be ready, my friend! He could come today!

Sunday Snippet Reflections (SSR)

1. How did today's Snippet speak into my "soul" garden?

2. Application to my "soul" garden.

3. One thing I am grateful for:

4. This week, I will commit to pray for:

5. Answered prayers:

Bible/Sermon Notes Date:___________

Scripture:

____

____

Key verses:

____

____

____

Notes:

____

____

____

____

____

____

____

____

____

____

____

____

____

____

ABOUT THE AUTHOR

Nancy Umberger is a writer, wife, and mom who holds a bachelor's degree in Christian Education and Psychology, a master's degree in biblical counseling, and certificates in life coaching. Her writing is inspired by her love for the Lord and her passion for being in nature. She is passionate about gardening, photography, cooking, hiking, her family, and above all, serving the Lord. Nancy resides in Winston-Salem, North Carolina, with her family.

To inquire about speaking at your ladies' event or Christian school for professional development, contact her at flashberger@att.net.

Made in the USA
Monee, IL
03 November 2026

67383025R00144